Physical Geography
A Self-Teaching Guide

Physical Geography
A Self-Teaching Guide

Michael Craghan

WILEY

John Wiley & Sons, Inc.

Published by John Wiley & Sons, Inc., Hoboken, New Jersey
Published simultaneously in Canada

For general information about our other products and services, please contact our Customer Care Department within the United States at (800) 762-2974, outside the United States at (317) 572-3993 or fax (317) 572-4002.

Wiley also publishes its books in a variety of electronic formats. Some content that appears in print may not be available in electronic books. For more information about Wiley products, visit our web site at www.wiley.com.

Library of Congress Cataloging-in-Publication Data:

Craghan, Michael
    Physical geography : a self-teaching guide / Michael Craghan.
        p.   cm.—(Wiley self-teaching guides)
    Includes index (p.).
    ISBN 978-0-471-44566-1 (pbk.)
    1. Physical geography.   I. Title.   II. Series.

GB59.C74 2003
910'.02—dc22   2003057677

10   9   8   7   6   5   4   3   2

Contents

Acknowledgments

Many thanks to Patricia Craghan, Elizabeth Maddalena, Pat Dunne, and my neighbors at 180 First for their good ideas and their faith. My appreciation goes to Allan Frei, Karen Nichols, Karl Nordstrom, Norbert Psuty, Dave Robinson, Michael Siegel, and other colleagues in Geography. Thank you to the people at John Wiley & Sons, especially Harper Coles, Jeff Golick, and Kimberly Monroe-Hill, who recognized the need for this book and encouraged me to work on it and brought forth just what I was imagining. Thank you also to Patricia Craghan, Andrew Craghan, Karen Caprara, N.W.U., and the Middle Atlantic Center for Geography and Environmental Studies for their assistance with photography and the production of this book. And to all of my family and friends: see, all those nights I really did go home to work on a book.

Introduction

Physical geography is the study of the forces that influence the surface of Earth. This book is intended to explain how geographic processes function and why they generate characteristic responses. Climate and geomorphology are the principal divisions in physical geography, and that is reflected in this book. The first part focuses on climatology, the study of atmospheric functions and their consequences. Some processes, such as the general circulation of the atmosphere, or the revolution of Earth around the Sun, are planetary in scale. Others, such as condensation or terrain effects on temperature, are more localized. The second part of the book is concerned with the solid earth. Geomorphology is the study of the processes that affect the surface of Earth and the landforms that are produced. Some of the processes are internal, such as plate tectonics, while others, such as stream flow, are external. Many geomorphic processes are driven by atmospheric or climatic forces. Always keep in mind that the surface of Earth and the atmosphere above it are constantly interacting with and influencing each other. Although each topical chapter may be studied in isolation, it is necessary to understand system linkages to fully appreciate environmental operations. At the end of each chapter I connect its themes with other sections in the book.

This book focuses on the aspects of physical geography that people are likely to encounter in their lives: the topics that pass the "Why should I care?" test—not the arcane elements or trivia. I have tried to select subjects that are prevalent or that are responsible for large proportions of system operations. Because of this book's purposes, topics are discussed at a very basic level, and I acknowledge that some things are greatly simplified. Readers should be aware that any subsection of these chapters would offer a lifetime of research opportunities to an Earth scientist. Because concepts, not details, are the foci of this work, its lessons should be applicable everywhere, although there is a bit of a North American bias to the presentation.

One of the appealing things about studying physical geography is its obvious relevance to society. When you consider the weather or climate, or when you read about a flood or an earthquake, you are thinking about how people are affected by environmental processes. Physical geography has real-world applications in fields such as disaster planning, agriculture, engineering, and environmental management. You will be able to open a good newspaper nearly every day and see how the topics in this book cross over into the social and political domains.

1 Earth and Sun

Objectives

In this chapter you will learn that:

- Earth is approximately 25,000 miles around.

- Earth rotates on its axis, which generates night and day.

- Latitude is an angle measurement used to identify a location on the surface of Earth.

- It takes Earth one year to revolve around the Sun.

- Seasons are caused by how the tilt of Earth's axis affects the orientation of the planet as it revolves around the Sun.

- Hours of daylight are determined by Earth's orientation with the Sun.

Size and Shape of Earth

Earth is a planet—it is a large body that moves around the Sun. It is not a perfect sphere, but Earth is a spherically shaped object. Earth has these approximate dimensions:

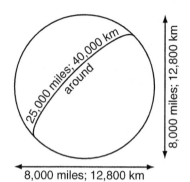

Figure 1.1. Earth is about 4,000 miles from its center to the surface (8,000-mile diameter) and approximately 25,000 miles around.

8,000 miles; 12,800 km

- Radius: 4,000 miles (6,400 km)

- Diameter: 8,000 miles (12,800 km)

- Circumference: 25,000 miles (40,000 km)

These values can vary slightly due to differences in surface topography and because Earth is not an exact sphere. If you could drive nonstop around the equator at 60 mph it would take seventeen days to make the trip.

What is the approximate distance around Earth (its circumference)?

_____

Answer: 25,000 miles (40,000 km)

Rotation, Poles, Equator

One feature of this planet is its rotation—it spins. It takes one day for Earth to rotate on its axis (one day exactly, because that is the definition of a day: one spin on its axis). Spinning leads to a reference system based on the axis of rotation. The North and South Poles are at the ends of the axis of rotation and thus can be used as unique reference points. If Earth did not spin (and thus had no rotation axis), then any place would be as good as any other for describing location.

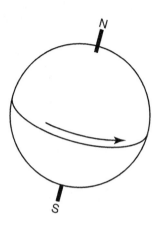

Figure 1.2. Because Earth rotates, we can identify the North Pole and the South Pole as special spots. A place on Earth will rotate once around and find itself back in the same position a day later.

Rotation also produces another feature of interest: the equator. The equator is in a plane perpendicular to the axis of rotation, and it divides the spherical Earth into halves. All of the points on one side of the equator are closer to the North Pole than to the South Pole. All of the points on the other side are closer to the South Pole. The half of Earth closest to the North Pole is called the Northern Hemisphere (half a sphere). The half of Earth closest to the South Pole is the Southern Hemisphere.

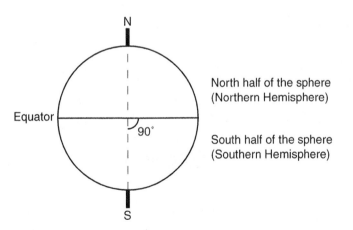

Figure 1.3. The equator is in a plane perpendicular to the axis of rotation, and it separates Earth into two halves: the Northern Hemisphere and the Southern Hemisphere.

What is the line that divides Earth into a half that is closer to the North Pole and another half that is closer to the South Pole? _____

Answer: the equator

Latitude

Once the two poles and the equator have been identified, then a system of measurement called latitude can be established. Latitude is an angle measurement from the equator to a point on Earth's surface. The angle is measured from the center of Earth at the point where the rotation axis intersects the plane of the equator.

The latitude system has some simple qualities:

- All points on the equator are 0° away from the equator.

- The North Pole is 90° away from the equator.

- The South Pole is 90° away from the equator.

- If the angle is measured toward the North Pole it is called north latitude.

- If the angle is measured toward the South Pole it is called south latitude.

- North and south are important! You must state whether a place has north or south latitude to properly identify it.

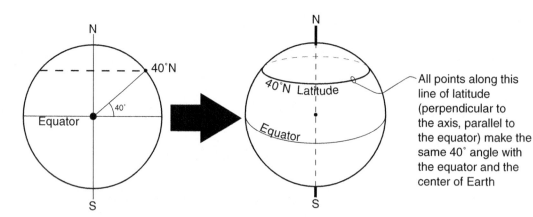

Figure 1.4. All points that are the same angle away from the equator and the center of Earth have the same latitude.

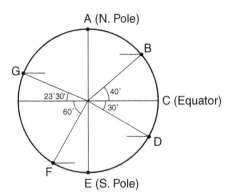

Figure 1.5.

What are the latitudes of points A, B, C, D, E, F, and G in Figure 1.5?

Answer: A = 90°N, B = 40°N, C = 0°, D = 30°S, E = 90°S, F = 60°S, G = 23°30'N

Revolution around the Sun

At the same time that it is rotating on its axis, Earth also is following a path around the Sun. Earth is a planet that rotates on its axis and also revolves around the Sun.

$$\text{Rotate} = \text{axis} = 1 \text{ day}$$

$$\text{Revolve} = \text{orbit} = 1 \text{ year}$$

It takes one year for Earth to revolve around the Sun (one year exactly, because that is the definition of a year: one trip around the Sun). This journey also takes 365¼ days (i.e., one year). So Earth will rotate on its axis 365¼ times in the time it takes for the planet to go around the Sun and return to its departing point.

The path that Earth travels along is an ellipse—but it is very close to being a circle. The nearly circular path is used to define a geometric feature called the plane of revolution. Although the planet orbits within the plane of revolution—this is going to affect almost everything on Earth— Earth's axis of rotation (the line running from the South Pole through the North Pole) always points toward the North Star.

Figure 1.6. It takes Earth one year to complete its nearly circular revolution around the Sun. Earth's axis is always tilted toward the North Star.

For the North Pole to be continuously directed toward the North Star, Earth's axis has to be tilted 23½° away from perpendicular to its plane of revolution around the Sun. The direction and angle of the tilt will always be the same: the axis is always aligned toward the North Star. As a result of its constant aim to the North Star, the alignment of the axis with the Sun is always changing. For part of its revolution around the Sun, Earth's North Pole generally leans toward the Sun, and for the other part of a year it leans away from the Sun.

- In December, the North Pole leans away from the Sun.

- In June, the North Pole leans toward the Sun.

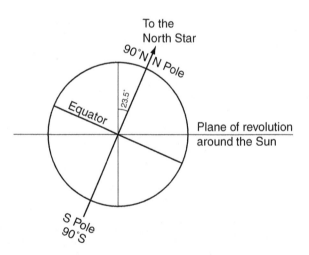

Figure 1.7. Earth's axis is tilted 23½° away from perpendicular to its orbit in the plane of revolution.

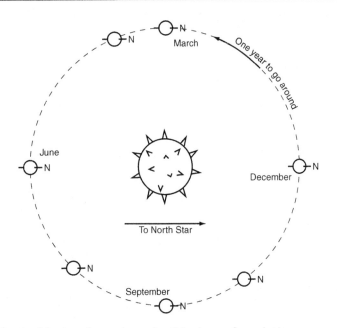

Figure 1.8. In this view from above Earth's plane of revolution you can see that the North Pole is always pointed toward the North Star. This causes the orientation of Earth with respect to the Sun to always be changing.

- In March and September, the line from Earth to the Sun is perpendicular to the South Pole–North Pole axis.

Because the North Pole is always pointing to the North Star, Earth's Northern Hemisphere is directed _____ the Sun in June and _____ the Sun in December.

Answer: toward; away from

Tilt and Reference Latitudes

This tilt of Earth's axis creates five special latitude lines. These five lines are the equator, two "tropics," and two "circles." Because tropics and circles are lines of latitude, they are in planes that are perpendicular to Earth's rotation axis and parallel to the plane of the equator.

Tropics are located at 23½°N and 23½°S, and just touch the plane of revolution around the Sun. Circles are located at 66½°N and 66½°S (23½° + 66½° = 90°), and they just touch the line that passes through Earth's center and is perpendicular to the plane of revolution. The areas bounded by these five latitude lines (equator, two tropics, and two circles) react in different ways to the changing orientation of Earth and the Sun over the course of a year.

- 66½°N is the Arctic Circle. As Earth rotates on its axis, all of the places on the North Pole side of this line will always be on the same side of perpendicular as the North Pole.

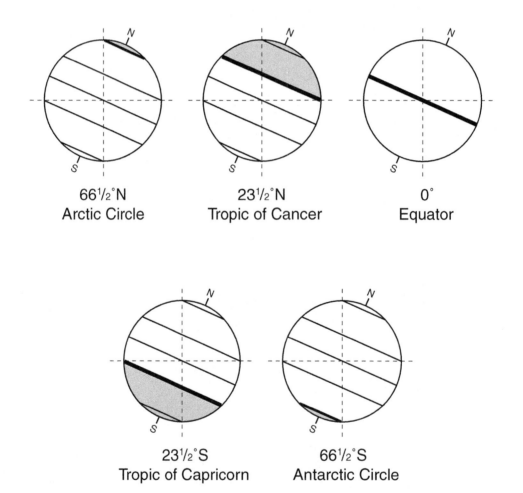

Figure 1.9.   There are five special lines of latitude that are produced by Earth's 23½° angle of tilt to its plane of revolution around the Sun.

- 23½°N is the tropic of Cancer. As Earth rotates on its axis, all of the places on the North Pole side of this line will always be above the plane of revolution around the Sun. No point that is north of this line will ever rotate to be directly on the plane of revolution.

- The equator is at 0° latitude. As Earth rotates on its axis, all points at the equator will spend half of each day above the plane of revolution and half below it, and half of each day on the North Pole side of perpendicular and half on the South Pole side.

- 23½°S is the tropic of Capricorn. As Earth rotates on its axis, all of the places on the South Pole side of this line will always be below the plane of revolution around the Sun. No point that is south of this line will ever rotate to be directly on the plane of revolution.

- 66½°S is the Antarctic Circle. As Earth rotates on its axis, all of the places on the South Pole side of this line will always be on the same side of perpendicular as the South Pole.

Which two lines mark the farthest places north and south that can be directly on Earth's plane of revolution around the Sun? _____

Answer: the tropic of Cancer (23½°N) and the tropic of Capricorn (23½°S)

Revolution, Alignment, and Day Length

As Earth travels around the Sun, Earth's tilt toward the North Star will create four days when Earth–Sun alignment is in a special condition. In June and December, there are solstices. A solstice is the moment when the Sun is directly overhead at one of the tropics. This is the farthest point north or south of the equator that the Sun can be directly overhead. A solstice also is the day when a hemisphere is aimed either most directly toward the Sun (summer solstice) or away from the Sun (winter solstice). In September and March, there are equinoxes. An equinox is the moment when the Sun is directly over the equator. Solstices and equinoxes mark the extremes of orientation and a changeover with respect to Sun conditions.

June Solstice

On the day of the June solstice, the North Pole is tilted as close as it gets toward the Sun and the South Pole is tilted as far away as it gets. It is summer in the Northern Hemisphere and winter in the Southern Hemisphere. On this day:

- The Sun will be directly overhead at 23½°N (tropic of Cancer), and it is strongest at that latitude.
- All points in the Northern Hemisphere will get more than 12 hours of sunlight; they spend more than half of the day rotating on the sunlit side of the planet. All points in the Southern Hemisphere will get fewer than 12 hours of sunlight.
- All points on the equator will spend 12 hours rotating on the sunlit side of Earth and 12 hours rotating on the dark side of Earth.
- All points north of the Arctic Circle (66½°N) will spend the entire 24-hour day rotating on the sunlit side of Earth.
- All points south of the Antarctic Circle (66½°S) will spend the entire 24-hour day rotating on the dark side of Earth.

September Equinox

On the day of the September equinox, the Sun is directly overhead at the equator. It is the first day of autumn in the Northern Hemisphere and the first day of spring in the Southern Hemisphere. On this day:

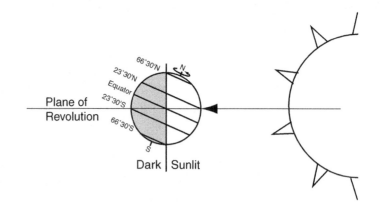

Figure 1.10. The June solstice.

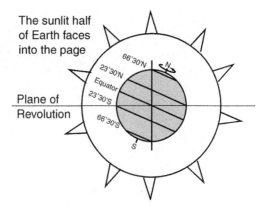

Figure 1.11.   The September equinox.

- The Sun is most directly overhead at the equator.

- All points on Earth will rotate on the sunlit side of the planet for 12 hours and rotate on the side away from the Sun for 12 hours.

December Solstice

On the day of the December solstice, the South Pole is tilted as close as it gets toward the Sun and the North Pole is tilted as far away as it gets. It is winter in the Northern Hemisphere and summer in the Southern Hemisphere. On this day:

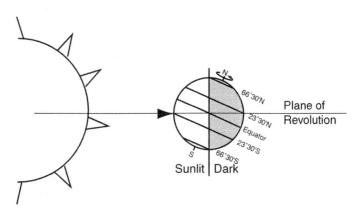

Figure 1.12.   The December solstice.

This classic photograph from December 7, 1972, was taken by a crew member on *Apollo 17* near the time of the December solstice. Note how the part of Earth that is sunlit and visible to the astronauts ranges from nearly all of Antarctica (at bottom) to the Mediterranean Sea (top) at about 40°N latitude. (Image AS17-148-22721 courtesy of Earth Sciences and Image Analysis Laboratory, NASA Johnson Space Center.)

- The Sun will be directly overhead at 23½°S (tropic of Capricorn), and it is strongest at that latitude.

- All points in the Southern Hemisphere will get more than 12 hours of sunlight; they spend more than half of the day rotating on the sunlit side of the planet. All points in the Northern Hemisphere will get fewer than 12 hours of sunlight.

- All points on the equator will rotate 12 hours on the sunlit side of Earth and rotate 12 hours on the dark side of Earth.

- All points south of the Antarctic Circle (66½°S) will spend the entire 24-hour day on the sunlit side of Earth.

- All points north of the Arctic Circle (66½°N) will spend the entire 24-hour day rotating on the dark side of Earth.

March Equinox

On the day of the March equinox, the Sun is directly overhead at the equator. It is the first day of spring in the Northern Hemisphere and the first day of autumn in the Southern Hemisphere. On this day:

- The Sun is most directly overhead at the equator.

- All points on Earth will be on the sunlit side of the planet for 12 hours and on the dark side for 12 hours.

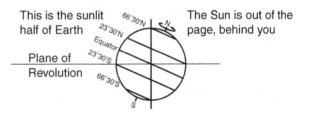

Figure 1.13. The March equinox.

"In-Between" Days

Because of the way Earth is tilted with respect to its plane of revolution, the Sun can never be directly overhead north of 23½°N (tropic of Cancer) or south of 23½°S (tropic of Capricorn). The Sun can only be overhead in the tropics (between Cancer and Capricorn). The Sun will be directly overhead at the equator on the two equinox days of the year (in March and September). Days that are not an equinox or a

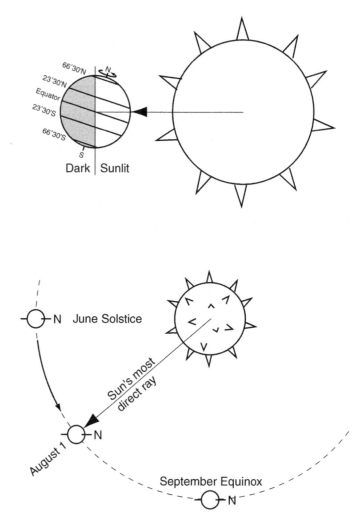

Figure 1.14. On August 1, the Earth–Sun relationship will be "in between" the conditions from the June solstice and the September equinox. *Top:* The Sun will be most directly overhead in the Northern Hemisphere, somewhere between the tropic of Cancer and the equator (it will actually be at about 18°N). On this day, all places in the Northern Hemisphere spend more than half a day on the sunlit side of Earth. *Bottom:* Don't be misled by two-dimensional depictions of the situation. The Sun is directly overhead at 18°N latitude. Earth is still tilted 23½° away from perpendicular with respect to its plane of revolution around the Sun. The North Pole is still aimed at the North Star.

solstice (i.e., the other 361) are simply "in the middle." If you interpolate between the solstice and equinox extremes, you should be able to figure out Earth–Sun relations for any day. Here are the basic principles:

- The Sun must be overhead somewhere between the tropics of Cancer and Capricorn.

    If it is March–September, the Sun will be directly overhead a place in the Northern Hemisphere.

    If it is September–March, the Sun will be directly overhead a place in the Southern Hemisphere.

    The closer a date is to a solstice, the closer to a tropic the Sun will be overhead.

    The closer a date is to an equinox, the closer to the equator the Sun will be overhead.

- Hours of daylight will be controlled by which hemisphere the Sun is "in" (i.e., in which hemisphere it is overhead) and then by latitude.

    In a hemisphere in which the Sun is overhead, the closer to a pole a place is in that hemisphere, the more hours of daylight there will be at that place.

    There are 12 hours of daylight every day of the year at the equator.

    In a hemisphere in which the Sun is not overhead, the closer to a pole a place is in that hemisphere, the fewer hours of daylight there will be at that place.

Why would a place like New York City (lat. 41°N) have about 15 hours of daylight in June but only about 9 hours in December? _____

Answer: The Northern Hemisphere is tilted most directly toward the Sun in June, and a place at 41°N spends most of a day on the illuminated part of Earth. In December, the Northern Hemisphere is tilted away from the Sun, and a place at 41°N spends most of a day on the dark side of Earth.

Seasonal Changes

Seasons are produced because Earth revolves around the Sun with a tilted axis, which directs different parts of the planet toward or away from the Sun at different stages of the journey.

Seasons have nothing to do with how close Earth and the Sun are to each other. If proximity was the cause of seasons, then it would be the same season in the Northern and Southern Hemispheres on the same day.

Yearly Sunlight Variations because of Earth's Revolution and Tilt

	Month	Sun Overhead	Daylight at North Pole	Daylight in the 48 states	Daylight at Equator	Daylight in Australia	Daylight at South Pole
Solstice	June	23½°N	24 hrs	>12 hrs	12 hrs	<12 hrs	0 hrs
Equinox	September	0°	12 hrs	12 hrs	12 hrs	12 hrs	12 hrs
Solstice	December	23½°S	0 hrs	<12 hrs	12 hrs	>12 hrs	24 hrs
Equinox	March	0°	12 hrs	12 hrs	12 hrs	12 hrs	12 hrs
Solstice	June	23½°N	24 hrs	>12 hrs	12 hrs	<12 hrs	0 hrs
	August 1	18°N	24 hrs	>12 hrs	12 hrs	<12 hrs	0 hrs
Equinox	September	0°	12 hrs	12 hrs	12 hrs	12 hrs	12 hrs
Solstice	December	23½°S	0 hrs	<12 hrs	12 hrs	>12 hrs	24 hrs
Equinox	March	0°	12 hrs	12 hrs	12 hrs	12 hrs	12 hrs
Solstice	June	23½°N	24 hrs	>12 hrs	12 hrs	<12 hrs	0 hrs

What causes seasons? _____

Answer: The tilt of Earth's axis changes the alignments of the two hemispheres and the Sun over the course of a year.

SELF-TEST

1. How much is Earth's rotation axis tilted away from perpendicular to the planet's plane of revolution around the Sun?
   a. 0°   c. 78½°
   b. 23½°   d. 90°

2. The moment when the Sun is directly over either of the two tropics is called a(n) _____.
   a. solstice   c. season
   b. equinox   d. ellipse

3. The Sun is directly above the tropic of Cancer (23½°N) in which of these months?
   a. February
   c. June
   b. April
   d. August

4. Earth's North Pole is always pointed toward the _____.

5. The Sun is directly above the equator in the months of _____ and _____.

6. The reason that it is hot in the summer is because that is when Earth is closest to the Sun. (True or False)

7. There are 12 hours of daylight at the equator every day of the year. (True or False)

8. How does Earth's rotation on its axis cause night and day?

9. With respect to Earth–Sun relationships, how are tropical latitudes different from middle and high latitudes?

10. For February 1, describe in general terms the latitude zone where the Sun will be directly overhead.

ANSWERS

1. b
2. a
3. c
4. North Star

5. March; September
6. False
7. True

8. Rotation spins a place into the sunlit half of Earth, then around out of the sunlight to the dark side of the planet.

9. The Sun can be directly overhead in tropical areas (between 23½°N and 23½°S), but it can never be 90° overhead in the middle or high latitudes.

10. The Sun will be directly overhead in the Southern Hemisphere, between the equator and the tropic of Capricorn.

Links to Other Chapters

- Latitude affects the length of day and the duration of daily insolation (chapter 2), which affects heating and temperature (chapter 2).
- Earth's rotation on its axis will produce night and a day with changing sunlight intensity, which will affect daily temperature patterns (chapter 2).
- Variations in heating help establish the general circulation of the atmosphere (chapter 5).
- Earth's spin on its axis contributes a Coriolis force, which affects how the wind moves across the surface of Earth (chapters 4, 5).
- Seasonal variations in heating are a major factor in climate (chapter 7).
- Temperature and seasons affect geomorphic forces such as weathering (chapter 11), soil formation (chapter 11), and glacial processes (chapter 14).

2 Insolation and Temperature

Objectives

In this chapter you will learn that:

- Solar energy does not hit all places on Earth with the same intensity.

- The more intense the Sun's rays are, the more energy the ground will absorb and the warmer it will be.

- Earth radiates as much energy away to space as it gets from the Sun.

- January and July usually are the coldest and hottest months, respectively, even though insolation is least intense in December and most intense in June in the Northern Hemisphere.

- Solar energy does not hit a place with the same intensity all day long.

- The coldest time of a typical 24-hour day is a bit after the Sun rises. The hottest time of a typical day is midafternoon.

- Continental places heat up faster and cool off quicker than coastal places.

Insolation Intensity

Energy that comes from the Sun to Earth is called *incoming solar radiation,* insolation. Insolation intensity changes at a given place from day to day during a year because the Earth–Sun orientation changes as Earth revolves around the Sun. Within a day, insolation intensity changes from minute to minute because a place's solar alignment is changing as Earth rotates on its axis. The Sun is stronger at noon than at early morning, and it has no power at all after sunset. At any given moment, different places on Earth will have different insolation intensities depending on (1) latitude or (2) time of day. Each of these two factors influences how directly the Sun will shine onto a place.

What two actions of planet Earth affect the amount of insolation being received at a given place? _____

Answer: The rotation of Earth on its axis and the revolution of Earth around the Sun each change the alignment of a place with respect to incoming solar radiation.

Seasonal Changes to Insolation

The place where the Sun is most directly overhead will receive more insolation energy than other places. Over the course of a year, the place where the Sun is directly overhead is changing every day. The Sun is directly overhead at the tropic of Cancer (23½°N) at the June solstice. The Sun is directly overhead at the tropic of Capricorn (23½°S) at the December solstice. At the March and September equinoxes, the Sun is directly overhead at the equator.

At times between solstices and equinoxes, the Sun will be overhead in tropical areas, the zone between 23½°N (tropic of Cancer) and 23½°S (tropic of Capricorn). The Sun will pass directly over a tropical place twice a year: once as the Sun moves from being overhead at the equator to being overhead at the tropic and then again on the return from the tropic to the equator. The Sun will never be overhead outside of the

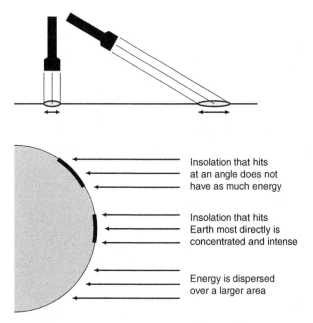

Figure 2.1. *Top:* A flashlight shining directly down onto a floor will have all of its light energy concentrated in a small, bright circle. If the flashlight is directed at an angle to the floor, the energy will be distributed over a larger dimmer area. Insolation is similarly concentrated or dispersed depending on the angle that it strikes the surface. *Bottom:* Insolation that reaches Earth at a high angle is concentrated and intense. As the angle of insolation decreases, the sunlight is dispersed and its intensity drops.

tropics; the Sun will be most overhead in middle and high latitudes on the day of that hemisphere's summer solstice.

When insolation reaches the surface of Earth, some of that solar energy is absorbed and turned into heat. Heating and warming from the Sun will vary over the course of a year as insolation intensity changes. The yearly changes in insolation intensity cause the temperature patterns we associate with seasons. It is hotter in summer because insolation is more direct and because there are more hours of daylight (a lot of high-intensity sunlight). It is cold in winter because insolation comes at a lower angle and there are fewer hours of daylight (little, weak sunshine).

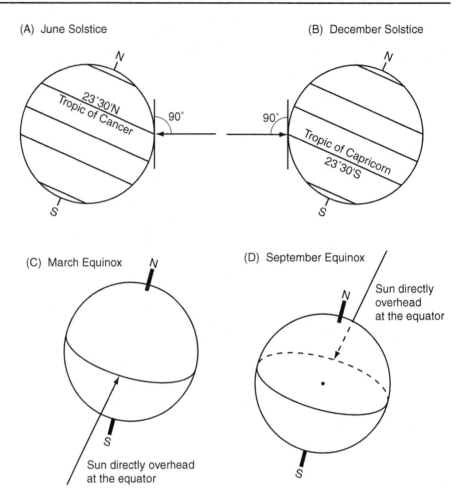

(A) June Solstice

(B) December Solstice

23°30'N
Tropic of Cancer

90°

90°

Tropic of Capricorn
23°30'S

(C) March Equinox

(D) September Equinox

Sun directly
overhead
at the equator

Sun directly overhead
at the equator

Figure 2.2. *A:* On the day of the June solstice, insolation is straight overhead at the tropic of Cancer, 23½°N. *B:* On the day of the December solstice, insolation is straight overhead at the tropic of Capricorn, 23½°S. At the *(C)* March equinox and the *(D)* September equinox, the Sun is straight overhead at the equator. Insolation intensity and therefore heating intensity will be greatest when the Sun is most directly overhead.

In what month will New York City (41°N) receive the most intense insolation? _____

Answer: June

Heating, Reradiation, and Yearly Temperature Cycles

At the same time that the Sun's energy is coming into the half of the planet that is in daylight, heat energy is being radiated away from the planet to space. There is a balance between incoming solar radiation and outgoing heat radiation. If Earth received more insolation than it radiated away, it would get hotter and hotter. If Earth received less insolation than it radiated away, it would get colder and colder.

There is a planetwide energy balance, but at any given moment a particular place can be out of equilibrium. The planetary radiation balance is not evenly distributed. When the Sun is high overhead (in summer), a place receives more insolation energy than it loses from radiating heat away, so it gets progressively hotter. When the Sun is weak and less direct (in winter), a place radiates away more energy than it receives as insolation, so it gets progressively colder.

Over the course of a year, a place will have different insolation inputs that will affect how hot or cold it will be. Let's consider the yearly temperature changes for a midlatitude, Northern Hemisphere city somewhere in the lower forty-eight United States.

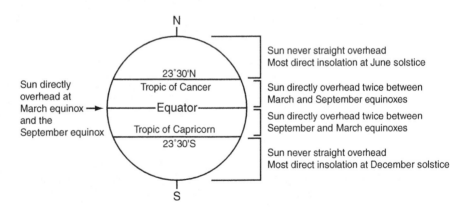

Figure 2.3. Insolation intensity is strongest when the Sun is directly overhead. In the tropics, the Sun will be directly overhead twice each year. In the middle and high latitudes, the Sun can never be directly overhead, but it is highest in the sky on the day of the summer solstice.

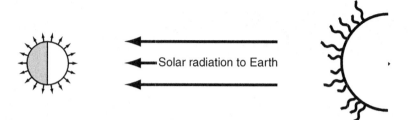

Figure 2.4. Earth continuously receives energy from the Sun—it is always daylight on half of the planet. Earth also gives energy off into space (in the form of invisible heat energy). All areas of Earth radiate energy, both the illuminated and the dark parts.

1. At the solstice in December, insolation will be at its lowest level. The Sun will be directly over the tropic of Capricorn in the Southern Hemisphere, and the Northern Hemisphere will be tipped as far away from the Sun as it gets. The angle of the incoming sunlight will be low, and the duration of daylight will be at its shortest. Our city will be receiving the least amount of solar energy and it will be losing radiated heat energy to outer space, so the temperature will be cold.

2. In January, insolation is slightly higher although still low. The Sun is not heating the city enough to make up for heat lost to radiation. Even though the insolation is getting stronger, it is still relatively weak, and the city is still getting colder. The average yearly temperature will be at its minimum in January.

3. From February to May, to our city's inhabitants, the Sun is getting progressively stronger. The length of daylight increases daily, and the Sun gets higher and higher in the sky every day. The heating of the Sun is exceeding the cooling from radiation losses, and the city gets a little warmer every week.

4. At the time of the June solstice, the Sun is overhead at the tropic of Cancer (its farthest point north), and the hours of daylight are at their maximum. There is a tremendous amount of insolation, because of the long days and the high Sun angle. The heating from the Sun greatly exceeds the amount of heat lost to radiation, so temperatures are increasing.

5. In July, insolation is still very high, but it's a bit weaker than it was at the solstice. Solar inputs are greater than radiation losses, so the

Yearly Insolation and Temperature Changes for a Midlatitude City in the
Northern Hemisphere

Month	Sun Overhead	Insolation	Temperature	Explanation
December	Tropic of Capricorn	Least	Cold	Sun is overhead deep in the Southern Hemisphere so there are a few hours of weak sunlight.
January	Southern Hemisphere	Very low	Coldest	Even though insolation is increasing, more heat is being lost from radiation.
February to May	Moving north	Increasing daily	Warming up	Increasing strength of the Sun over-comes heat losses from radiation.
June	Tropic of Cancer	Most	Hot	The Sun is high overhead, and there are long days of strong sunlight.
July	Northern Hemisphere	Very high	Hottest	Insolation is still adding more heat than is being radiated away.
August to November	Moving south	Decreasing daily	Cooling down	Decreasing strength of the Sun can't replace the heat that is radiated away.
December	Tropic of Capricorn	Least	Cold	Sun is overhead deep in the Southern Hemisphere so there are a few hours of weak sunlight.

temperature is still increasing. July usually has the hottest average monthly temperature of a year.

6. From August to November, the Sun gets progressively weaker at this city. The point of maximum insolation is moving southward and will enter the Southern Hemisphere after the September equinox. The duration of daylight is decreasing, and the insolation

angle also is dropping. Daily energy inputs are dropping below radiation losses, and it is getting progressively cooler.

Why is January usually the coldest month of the year in mid- and high-latitude places in the Northern Hemisphere? _____

Answer: In January, the Sun is at a low angle, and therefore insolation strength is weak and the duration of the daylight is short. Even though the Sun is heating things more than it did in December, it is not adding enough heat to make up for the loss of energy out to space. By February, the increasing strength of the Sun can begin to add more heat than is lost.

Heating and Daily Temperature Cycles

The temperature pattern over a single day follows a pattern that is similar to the yearly changes. The pattern is a function of outgoing heat radiation and the minute-to-minute changes in the strength of incoming sunlight. Consider the example of a day when the Sun rises at about 6:00 A.M. and sets at about 6:00 P.M. over our city.

1. From 6:00 P.M. the night before until the sun rises at 6:00 A.M., there is no sunlight. During this time, the city is radiating away heat and not receiving any insolation, so it gets progressively colder from sunset until the next morning.

2. At 6:00 A.M., the Sun comes up and begins providing insolation. The Sun is very weak at this moment (it is very low in the sky), so the insolation still is less than the radiation being lost. It is still getting colder even though the Sun is up.

3. At some point a bit after sunrise, the increasing strength of the sunlight will equal what is being lost in reradiation. This is the coldest part of the day—after the sun is up (this is usually when there is dew, fog, or frost).

4. For the rest of the morning, the sunlight will get stronger and keep warming things up (it gets warmer, the dew or frost goes away, and the fog "burns off"). The sunlight keeps getting stronger—warming things up—and maximum insolation will occur at noon.

5. Even though the sunlight will be strongest at noon, this is *not* the hottest time of day.

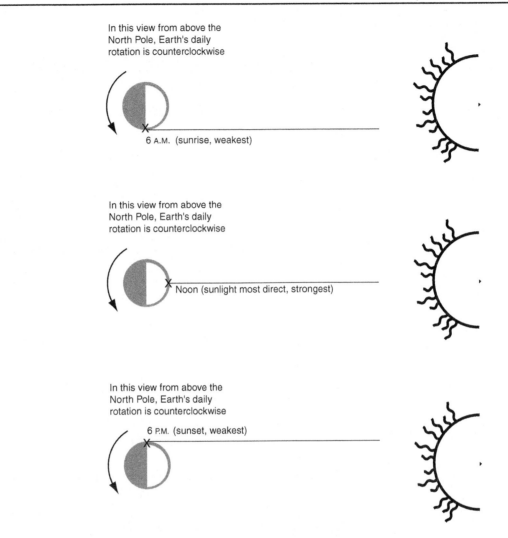

Figure 2.5. *Top:* At 6:00 A.M. the Sun comes up, and insolation begins to increase from its overnight strength of zero. *Middle:* At noon, the Sun is most directly overhead, and insolation (although not temperature) will be at its maximum for the day. *Bottom:* As the Sun sets, the strength of insolation drops to zero, and it will remain at zero until sunlight begins to illuminate a place at dawn.

6. In the afternoon, the sunlight is still strong, and it will keep heating up Earth.

7. At about midafternoon, the temperature difference between the lowest part of the atmosphere, which is closest to the warmed-up ground, and the higher parts of the atmosphere becomes strong,

Typical Daily Insolation and Temperature Changes

Time of Day	Insolation	Temperature Situation
Sunset to sunrise	None	Radiating heat, getting cooler
Sunrise	Onset	Sun weak, still losing heat
A bit after sunrise	Weak but picking up	Insolation = radiation; it will start warming
Morning	Increasing	Getting warmer
Noon	Maximum	Getting warmer
Early afternoon	Strong but decreasing	Still adding heat, getting warmer
Midafternoon	Losing strength	Ground-level air rises and replaced by cooler air; maximum temperature was reached
Late afternoon	Fading away	Weaker sun + new cool air: not able to get back up to the highest temperature
Sunset to sunrise	None	No insolation, getting progressively cooler

and the ground-level air begins to rise—hot air rises. The hottest part of the day has now been attained and passed. The air that moves in to replace the warm air that rose away will be colder and won't have all day to heat up. (Rising afternoon air helps produce late-day thunderstorms.)

8.   As the afternoon wears on, heat is being radiated away, and the sunlight keeps getting weaker and weaker until the Sun goes down.

The daily temperature cycle is superimposed on top of the yearly temperature cycle. In midlatitude places in the Northern Hemisphere, January mornings are likely to be the coldest times in winter, while July afternoons are likely to be the hottest times of the year.

Why is it usually hotter in the middle of the afternoon than when the sunlight is strongest at noon? _____

Answer: Even though the sunlight's strength peaks at noon, it is still strong enough to add heat to a place into the afternoon.

Continentality and Altitude

Continentality is an indication of how far from a large body of water a place is. This will influence how hot or cold a place gets. It takes a greater quantity of heat to warm up water in the ocean than it does to warm up the rocks and soil of the landmasses. Therefore land heats up much faster than water does. Water has a much greater ability to hold heat than the land. So once the ocean does warm up, it will stay warmer for longer periods than the land.

The moderating effects of a large body of water make it much harder to change the temperature of a coastal place than a continental place. Areas adjacent to large water bodies will not get as hot as inland places in the summer because the water will be colder than the land. Coastal areas will not get as cold as inland places in the winter because the ocean will be warmer than the land. Continental places usually have a larger temperature range on an annual basis (hotter summers and colder winters) and on a daily basis (hotter daily high temperatures and colder daily low temperatures) than comparable coastal places will.

In this June 6, 1991, photograph from the space shuttle, clouds cover the East Coast of the United States from New Jersey *(top)* to North Carolina *(bottom)*. The convection clouds are over the land, which in June would be much warmer than water bodies such as Delaware Bay, Chesapeake Bay, Pamlico Sound, and the coastal Atlantic Ocean, which are cloud-free in this image. (Image STS040-77-20 courtesy of Earth Sciences and Image Analysis Laboratory, NASA Johnson Space Center.)

If a body of water freezes, it is no longer able to influence the climate the way that liquid water can. (Sunlight can penetrate water; currents can mix heat through water; water has a higher specific heat than ice.) If a large lake freezes over in winter, it will not be able to moderate temperatures along its shoreline, as it can when it is liquid water.

Elevation also affects temperature. Air is thinner as altitude increases. If there are fewer atmospheric particles, there is less ability to

The higher one gets in the troposphere, the less dense the atmospheric gases, and the colder the air. This 1925 photograph shows a National Advisory Committee for Aeronautics pilot who is wearing a fur-lined leather flying suit with an oxygen facepiece for a test flight in this open-cockpit biplane. (Photograph GPN-2000-001380 courtesy of NASA.)

hold heat. It is almost always colder at higher elevations than at nearby lower places.

How would the yearly temperature range at a midlatitude place in the center of a continent (e.g., Lincoln, Nebraska) differ from that found at a coastal city at the same latitude? _____

Answer: Continental Lincoln is likely to have hotter temperatures in summer and colder temperatures in winter.

SELF-TEST

1. At what time of day is insolation the most intense?
   a. a bit after sunrise   c. at noon
   b. one hour before noon   d. in midafternoon

2. In a tropical place such as Ciudad Bolívar in Venezuela (lat. 8°N), when is the Sun precisely overhead?
   a. never   c. April and September
   b. at the June solstice   d. November and January

3. A _____ place will have greater annual temperature extremes than a coastal place at the same latitude.

4. A place receives _____ insolation between sunset and sunrise.

5. _____ is the time when a typical day has the hottest temperature.

6. The coldest time of a typical day is at midnight. (True or False)

7. Once the June solstice passes, a Northern Hemisphere city such as Lincoln, Nebraska (lat. 41°N) receives less and less insolation every day until the December solstice. (True or False)

8. How can insolation strength change over the course of a year at the equator if every day has 12 hours of sunlight?

9. Why can a large body of water affect the temperature of nearby cities?

10. Earth is continuously receiving light and heat energy from the Sun. Why doesn't Earth keep getting hotter and hotter over time?

1. c 2. c 3. continental or inland 4. zero or no

5. Midafternoon 6. False 7. True

8. Insolation strength will change with the angle that the sunlight hits the ground. The Sun is 90° overhead at the equator at the two equinoxes; it strikes at a lesser angle every other day of the year.

9. Water has a higher specific heat, which means it heats up slower and cools down slower than earth and rock. Also, water is semitransparent, so sunlight can pass through it to transmit heat below the surface. In addition, currents can distribute heat throughout the water. Because a large body of water can have a different temperature, it will influence the climate of nearby places.

10. Because the amount of energy received from insolation is balanced by the amount of heat energy radiated back into space.

Links to Other Chapters

- Earth–Sun alignments are the principal control on insolation intensity (chapter 1).
- Latitude affects how directly insolation will strike a place (chapter 1).
- There are fewer air particles at high altitudes than at lower ones (chapter 3).
- The hotter the air, the more moisture it can hold (chapter 3).
- The intertropical convergence zone will be found near the part of Earth with the greatest insolation and heating (chapter 5).
- Seasons, continentality, and altitude will affect a place's temperature and therefore its climate (chapter 7).
- Physical weathering from plants and ice is influenced by temperature. The speed of chemical weathering reactions is affected by temperature (chapter 11).
- The formation of glaciers requires an annual temperature that permits a yearly buildup of snow (chapter 14).

3 The Atmosphere and Atmospheric Water

Objectives

In this chapter you will learn that:

- The atmosphere is mostly made up of gases and has layers; the troposphere is the lowest layer.

- When insolation reaches Earth, it first encounters the atmosphere, which affects how it proceeds.

- UV radiation from the Sun is absorbed by ozone in the stratosphere.

- Because higher in the troposphere there are fewer and fewer particles that can hold heat, temperatures decrease with increasing altitude.

- The gas particles in the atmosphere exert a force called air pressure.

- When liquid water absorbs energy, the water can evaporate, becoming water vapor gas in the atmosphere.

- Relative humidity is the amount of water vapor being held by a parcel of air expressed as a percentage of the amount of water vapor that the air can possibly hold.

- When air loses its ability to hold as much water vapor as it already has, some of that water vapor gas must revert to liquid water drops.

- As temperature increases, the ability of the air to hold water vapor increases. As air temperature decreases, there is less ability to hold water vapor.

- Rising air cools and loses its ability to hold water vapor. Descending air warms and increases its capacity to hold water vapor.

Atmospheric Layers

The atmosphere is made up of gases that surround the surface of Earth. The atmosphere has many different layers that remain mostly separate from one another. While all the layers play some role in earthly or human affairs, two layers have highly significant impacts: the troposphere and the stratosphere. The layer closest to Earth is the *troposphere*. All human activity except high-altitude aviation and space flight takes place within this layer. Nearly all weather originates here. The next layer up is the *stratosphere*. We care about the stratosphere for two principal reasons. First, the jet stream can be found there. The jet stream helps push air from place to place, influencing the weather. Second, the stratosphere holds ozone gas, which protects Earth's surface from ultraviolet (UV) radiation. UV rays can cause sunburn, skin cancer, and cataracts.

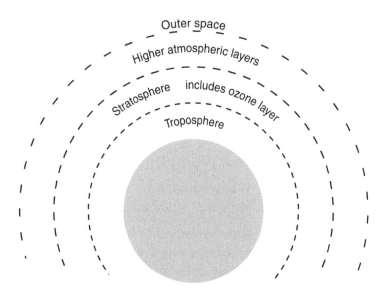

Figure 3.1. The atmosphere is made up of different gas layers that remain mostly separated from one another.

Most of the atmospheric gases and particles are close to the surface of Earth. The farther (higher) one gets from the surface, the fewer gases and particles there will be. Eventually there will be none; this marks the start of outer space.

All significant human activity occurs in which layer of the atmosphere?
_____

Answer: the troposphere

Atmospheric Constituents

The atmosphere is made up of various gases and solid particles. Most of its constituents are gases.

- Nitrogen gas (N_2) makes up most of the atmosphere, 78 percent.

- Oxygen gas (O_2) makes up most of the rest, 21 percent.

- Everything else makes up about 1 percent.

- Carbon dioxide gas (CO_2) is about 0.04 percent and rising.

In addition to the gases present in the atmosphere, there are other particles:

dust	liquid water	soot
pollen	exhaust	ice
smoke	artificial gases	volcanic ash

When people refer to the "ozone layer," it is a reference to the stratosphere. When ultraviolet radiation from the Sun comes to Earth, much of it is absorbed by ozone (O_3) in the stratosphere. This protects the surface of Earth from getting too much radiation. The ozone layer has been of some concern because artificially manufactured gases called chlorofluorocarbons (CFCs) have ascended into the stratosphere. CFCs can cause a chemical reaction that destroys the protective ozone.

Another extremely important gas in the atmosphere is water vapor. Water vapor is water in a gaseous state. It is not steam—the atmosphere is not over 212°F (100°C). Steam also is water in a gaseous state, but this is not what we are referring to. A gas will expand to fill an area; it is not confined to a given shape, like a solid, or to a given volume, like a liquid. Water vapor in the atmosphere behaves as a gas—it spreads all over. You recognize water vapor in the atmosphere as dampness or humidity. Sometimes the water vapor changes from the gaseous state and becomes a liquid (clouds, fog, dew, condensation, rain).

Why is there great concern about the destruction of stratospheric ozone? _____

Answer: Ozone absorbs ultraviolet radiation from the Sun, so if the amount of ozone decreases, more harmful UV radiation can reach the surface of Earth.

Intercepting Insolation

When energy from the Sun approaches Earth, it first encounters the gases and other particles in the atmosphere. The atmosphere will affect the incoming solar radiation.

- Some of the insolation is scattered, or deflected, from its original path when it hits atmospheric constituents.

- Some of the insolation penetrates the atmosphere and reaches the surface of Earth.

- Some of the insolation is absorbed by atmospheric gases and particles, which then heat up.

- Some of the insolation is reflected back to outer space.

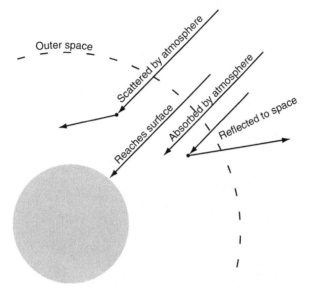

Figure 3.2. Some incoming energy from the Sun is intercepted by the particles and gases of the atmosphere.

How does the atmosphere protect Earth from the full strength of incoming solar radiation? _____

> *Answer:* Some of the radiation is scattered or deflected from its path by atmospheric particles, some radiation is absorbed by the atmosphere, and some is reflected back to space.

Temperature and the Atmosphere

Because heat can be absorbed and held by gas molecules or particulates in the atmosphere, the quantity of atmospheric materials affects the amount of heat that can be held. The fewer the heat-holding things, the colder it can be. In this way the decreasing amount of atmospheric material found with increasing elevation affects air temperature.

Air temperature in the troposphere (the lowest atmospheric layer) drops approximately 3.5°F with every 1,000-foot increase in elevation (or 6.4°C each kilometer). This environmental lapse rate is due to the decrease in the quantity of heat-holding atmospheric material with increasing altitude. This is why mountains are colder than lower elevations; why there can be snow in the mountains and not elsewhere; and why mountains can be snowcapped in summer, even in equatorial or tropical places.

At high elevations such as at the top of Mauna Loa, snow can fall, even in Hawaii. The loss of heat with altitude can overcome the otherwise warm tropical climate. (Photograph by John Bortniak/National Oceanic and Atmospheric Administration/Department of Commerce.)

Why does air temperature in the troposphere decrease with increasing altitude? _____

> *Answer:* As the distance above the surface increases, the number of heat-holding atmospheric particles decreases.

Air Pressure

The diminishing amount of atmospheric material with increasing elevation also affects air pressure. Air pressure is the force that the atmospheric particles put on other atmospheric particles and on Earth's surface. The more gas particles there are, the greater the pressure that the gas can exert (e.g., putting more air into a balloon increases the pressure the balloon is subject to, and can burst it).

The air in the atmosphere exerts a force in all directions, including onto the surface of Earth. We measure that force with a barometer ("bar" = pressure) and call it *atmospheric pressure*. The lower a place is, the more air particles there will be, and the greater the air pressure will be. Pressure will be higher at sea level than on top of a mountain. When people go upward quickly (in an airplane, on a hilly road, or in an elevator), their ears may pop to adjust for the lower atmospheric pressure. Lower air pressure at high altitudes also is why airplanes have pressurized cabins—so that people feel comfortable in what otherwise would be a situation of very low air pressure high in the sky.

What is atmospheric pressure? _____

Answer: It is the force that atmosphere gases put onto surrounding objects.

Evaporation

Liquid water becomes a gas in the atmosphere through evaporation. Evaporation happens when water absorbs enough energy to make the phase change from a liquid to a gas. The energy could be from the Sun's insolation, or from heat in the environment. Evaporation can happen at night; sunlight is not required.

If a container of water is left standing, eventually all of the water will disappear. Water in lakes, oceans, and soil can evaporate too. Anytime something that is wet dries off, or a pool of water dries up, the liquid water disappears due to evaporation. When liquid water evaporates, it

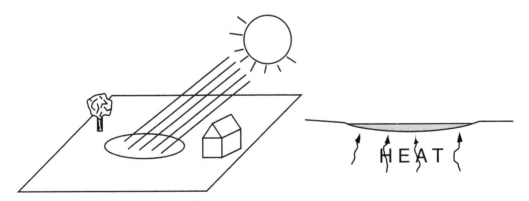

Figure 3.3.   *Left:* Sunlight can provide enough energy to evaporate water. *Right:* Water can disappear from a puddle if it absorbs enough energy from the environment to evaporate.

becomes water vapor (a gas) in the atmosphere, and is diffused throughout the environment. A liquid has a definite volume. A gas, such as water vapor, can expand into its surroundings; dampness or humidity can fill any space.

How does liquid water evaporate into the atmosphere?

_____

> *Answer:* Liquid water evaporates when it absorbs enough energy to become a gas.

Relative Humidity

The atmosphere has a limited ability to hold water vapor—there is a maximum load that is determined by air temperature.

• The *warmer* the air temperature is, the *more* water vapor it can hold.

• The *colder* the air temperature is, the *less* water vapor it can hold.

The amount of water vapor in the air is usually stated as a percent of the total water vapor that the air can hold. This percentage is called relative humidity (RH).

$$RH = \frac{\text{amount of water vapor present}}{\text{amount of water vapor possible}}$$

When the relative humidity is 100 percent, that means the air is full—it cannot handle any more water. It is already holding as much water vapor as it is capable of carrying. Air can never have more than 100 percent humidity; it can never have more water vapor than it is capable of holding.

Because the ability to hold water vapor depends on the temperature, the relative humidity will change if the temperature changes. If heat is taken away—if the temperature drops—then some of the capacity to hold water vapor also is taken away. The water vapor already present will take up a greater percent of the air's ability to hold water. If heat is added—if the temperature rises—then the capacity to hold water vapor also is increased. The water vapor already present will take up a smaller percent of the air's capacity to hold water vapor.

Let's look at some examples:

1. The air is 60°F and at 60 percent relative humidity. If the air and amount of water vapor remain the same but the temperature increases to 70°, then relative humidity will decrease because warmer air can hold more water vapor than it could before it warmed up.

$$RH = \frac{present}{possible}$$ If "possible" increases, then RH drops.

2. The air is 60°F and at 60 percent relative humidity. If the air and amount of water vapor stay the same but the temperature drops to 50°F, then relative humidity will increase. The same amount of water vapor will be there, but the ability of the air to hold it will decrease.

$$RH = \frac{present}{possible}$$ If "possible" decreases, then RH rises.

3. The air is 60°F and at 60 percent relative humidity. If the temperature stays the same but more water is evaporated into the air, then relative humidity will increase. The amount of water vapor will increase, and the ability of the air to hold the water vapor does not change.

$$RH = \frac{present}{possible}$$ If "present" increases, then RH rises.

- Temperature increases: relative humidity decreases.

- Temperature decreases: relative humidity increases.

- Relative humidity never exceeds 100 percent.

What does it mean if relative humidity is said to be 100 percent?

_____

Answer: The air is holding as much water vapor as it can carry.

Saturation and Condensation

Air that is holding as much water vapor as it possibly can (100 percent relative humidity) is said to be saturated. If saturated air gets colder, it will lose its ability to hold the water vapor it already contains. This means that some of the water vapor already present will be forced to become liquid water. When gaseous water vapor reverts to liquid water, it is said to *condense*. The "condensation" on a cold drinking glass, for example, is liquid water that came from water vapor in the air.

Liquid water in the atmosphere forms clouds or fog. If cloudy or foggy air warms up (gets more ability to hold water vapor), then that liquid water in the air can be reevaporated back into the gaseous state. This is why morning fog disappears as the Sun begins to warm things up.

On the other hand, if air temperature drops and keeps dropping:

1.   The ability of the air to hold water vapor drops, so relative humidity increases.

2.   Humidity eventually reaches 100 percent—the air becomes saturated with water vapor.

3.   The air can't hold as much water vapor as is already present, so some will be forced to condense back into liquid water.

4.   Clouds of liquid water will form in the atmosphere.

Once the air is saturated, any decrease in air temperature will force water to condense back into a liquid. If the air still keeps getting colder, then more vapor must revert to liquid. As gaseous water vapor condenses out into liquid water clouds, that can result in rain.

How do clouds of liquid water form in the atmosphere?

_____

Answer: Clouds form when saturated air loses some ability to hold water vapor, and therefore some of the vapor must condense back into a liquid state.

Adiabatic Heating and Cooling

Adiabatic processes are related to the physics of pressure, volume, and temperature of a gas (e.g., air). As a result, adiabatic processes affect air temperature and therefore affect the ability of the atmosphere to hold water vapor.

The two most important adiabatic consequences in geography are:

1. When air rises, it gets colder.

2. When air sinks, it gets warmer.

This is not the same concept in which there is an environmental lapse rate that corresponds to colder temperatures at higher altitudes (see "Temperature and the Atmosphere" earlier in this chapter). That temperature difference is caused by lower numbers of heat-holding particles at higher elevations. In that case, the air is not changing temperature; instead, there is air of different compositions at different altitudes. Adiabatic processes refer to a defined parcel of moving air going up or down. Changes to that parcel's volume or pressure will change its temperature.

If the temperature of a lifting parcel of air drops far enough, it can

Figure 3.4. *Top:* A parcel of air at the surface has a particular temperature and relative humidity. *Middle:* If the air is lifted up, its temperature drops so its relative humidity increases. *Bottom:* If the air continues to rise, its temperature may drop to the point where the air can no longer hold all the water vapor it has. The water vapor has to change back to liquid water—so a cloud will form.

On a beautiful summer day, convection can still send air up to its lifting condensation level, which will cause clouds to form. Note how the clouds are flat-bottomed and at approximately the same height. (Photograph courtesy of MACGES.)

become saturated as its ability to hold water vapor decreases. If the saturated air continues to rise and cool, then some of the water vapor must condense back into liquid water in the atmosphere (e.g., a cloud). The elevation where rising air becomes saturated and where clouds begin to form is called the lifting condensation level (LCL). If the liquid water droplets in a cloud get big enough, then gravity will be able to pull them down to the surface. It will rain (or snow).

When air descends, the opposite happens. Instead of cooling, descending air warms up, so its ability to hold water vapor increases.

Rising air is associated with clouds and rain or snow. There are five important processes that cause air to rise and cool off:

1. Air that converges at the surface has to go somewhere; it goes up.

2. Low-pressure systems have rising air near their centers.

3. Air that is warmer than its surroundings will rise by convection.

4. Cold and warm fronts will cause air to rise.

5. Air crossing some terrain may have to rise to pass obstacles.

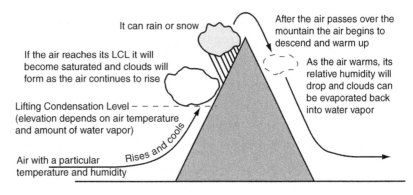

It can rain or snow

After the air passes over the mountain the air begins to descend and warm up

If the air reaches its LCL it will become saturated and clouds will form as the air continues to rise

As the air warms, its relative humidity will drop and clouds can be evaporated back into water vapor

Lifting Condensation Level – – – – (elevation depends on air temperature and amount of water vapor)

Rises and cools

Air with a particular temperature and humidity

Figure 3.5. Air approaches and moves over a mountain. The up-and-down movement of a parcel of air will adiabatically change the air temperature, which will affect the ability of the air to hold gaseous water vapor.

Sinking air is associated with blue skies and dry conditions. There are three important ways to get air to sink and warm up:

1. Converging winds in the upper troposphere have to go somewhere; they go down.

2. High-pressure systems have sinking air near their centers.

3. After passing over high terrain, the wind may follow the ground downward.

Why is rising air associated with clouds and precipitation?

_____

Answer: As air rises, it cools adiabatically and loses its ability to hold water vapor. If the air cools below its saturation temperature, the water vapor will condense back into liquid water and form clouds.

SELF-TEST

1. Oxygen and nitrogen gases together make up approximately
   _____ of the atmosphere.
   a. 10%   c. 50%
   b. 25%   d. 99%

2. Which atmospheric layer is adjacent to Earth's surface?
   a. troposphere   c. ozone layer
   b. stratosphere   d. outer space

3. The force that the atmosphere exerts on objects and particles it contacts is
   a. temperature   c. condensation
   b. humidity   d. air pressure

4. The gas in the stratosphere that absorbs UV radiation and keeps it from reaching the surface at full strength is _____.

5. When liquid water evaporates, it is transformed into a gas called _____.

6. Liquid water can evaporate only when it is exposed to sunlight. (True or False)

7. Saturated air has a relative humidity of 0%. (True or False)

8. How is relative humidity calculated?

9. Why is descending air associated with clear, blue skies?

10. How is the environmental lapse rate different from adiabatic temperature changes?

ANSWERS

1. d   2. a   3. d   4. ozone

5. water vapor   6. False   7. False

8. It is the amount of water vapor present in the atmosphere as a percent of the amount of water vapor the atmosphere can hold.

9. Descending air is heating adiabatically and increasing its ability to hold water vapor and therefore getting farther from being saturated.

10. The environmental lapse rate is associated with the decreasing number of atmospheric particles at higher altitudes. The adiabatic lapse rate is associated with changing properties of a particular moving parcel of air.

Links to Other Chapters

- Differences in air pressure from place to place will cause the atmosphere to be pushed from high-pressure places to areas with lower pressure (chapter 4).
- When air descends in high-pressure systems it produces dry, cloud-free conditions (chapter 4).
- When air rises in low-pressure systems it tends to produce precipitation (chapters 4 and 6).
- Low pressure at the intertropical convergence zone and subpolar low will lead to rainy conditions (chapter 5).
- Air does not readily mix between atmospheric layers. Air in the troposphere tends to remain there. This affects the general circulation of the atmosphere (chapter 5).
- The seasonal advance/retreat of high- and low-pressure elements of the general circulation of the atmosphere will produce yearly precipitation patterns (chapters 5 and 7).
- When air is lifted by warm or cold fronts it tends to produce precipitation (chapter 6).
- Gaseous water vapor will be the source for rain and snow (chapter 7).
- Decreasing temperature with increasing altitude will produce a special "highlands climate" that will be colder than surrounding lower places (chapter 7).
- Very cold air (e.g., near the poles) will be dry because it doesn't have the ability to hold much moisture (chapter 7).
- Weathering and soil-forming processes are very dependent on the availability of water (chapter 11).
- The position of the water table and the availability of groundwater are functions of precipitation (chapter 12).
- Stream floods are produced when there is more precipitation and runoff than can be handled within the channel (chapter 13).
- Glaciers are produced in places where more snow falls than melts from year to year (chapter 14).

4 Pressure and Wind

Objectives

In this chapter you will learn that:

- Atmospheric pressure varies from place to place.

- Air is pushed from places with high pressure to areas with lower atmospheric pressure.

- When air moves across the surface of Earth it is turned from its path by Coriolis force.

- Air rotates clockwise around high-pressure centers in the Northern Hemisphere.

- Air rotates counterclockwise around low-pressure centers in the Northern Hemisphere.

- Near the ground, air spirals clockwise and away from Northern Hemisphere high-pressure centers.

- Near the ground, air spirals counterclockwise and in toward Northern Hemisphere low-pressure centers.

- High pressure is characterized by descending air. Low pressure has rising air.

Air Pressure

The atmosphere is not uniform. There are different solids and gases in different places. Insolation intensity, and therefore temperature, will vary with latitude and cloud cover. Different air masses have different humidities. Because of the temperature, humidity, and constituent differences, in some places the atmosphere will press with more force than it does in other places. Places where the atmosphere has more force than average are said to have high pressure. Places where the atmosphere has less force than average are said to have low pressure.

Different places can have different atmospheric pressures. (True or False)

Answer: True

Isobars and Pressure Centers

On a map, atmospheric pressure is indicated with lines called *isobars* (iso = same, bar = pressure). Isobars connect places that have the same pressure; all places on an isobar line have the same atmospheric pressure. If all points on an isobar line have the same air pressure, that means that the atmosphere on one side of the line will have higher air pressure than the air on the other side.

A spot that is completely surrounded by places that have lower atmospheric pressure is called a high-pressure center. A high-pressure center (usually indicated on a weather map by "H") is surrounded by areas with decreased barometric pressure. Each successive isobar away from the high-pressure center represents *lower* air pressure. The isobars that are closer to the high-pressure center indicate higher atmospheric pressure than the isobars that are farther away.

A low-pressure center (usually indicated by "L") is surrounded by areas with increasing barometric pressure. Each successive isobar away from the low-pressure center represents *higher* air pressure. The isobars that are closer to the low-pressure center indicate lower atmospheric pressure than the isobars that are farther away.

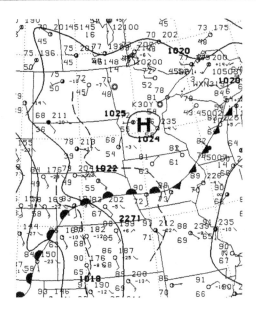

The wavy, solid lines on this weather map are isobars. There is a high-pressure center in Minnesota with a value of 1,025 millibars; it is surrounded by areas of lower atmospheric pressure. On this map isobars are shown in 4-millibar intervals (e.g., 1,024, 1,020, 1,016, etc.). The line marked with semicircles and triangles is a boundary between two air masses. (Weather map courtesy of National Weather Service.)

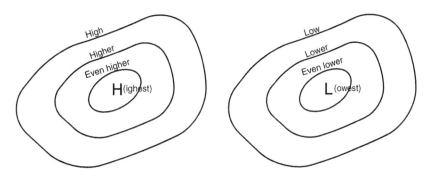

Figure 4.1. *Left:* Nested isobars around a high-pressure center represent higher air pressure the closer they are to the center. *Right:* Nested isobars around a low-pressure center represent lower air pressure the closer they are to the center.

A place that is surrounded by areas that have *higher* atmospheric pressure is called a _____.

Answer: low-pressure center

Pressure Force

It is almost never important what value of pressure a place has. What matters is how much atmospheric pressure it has compared to neighboring places. Air is pushed from places with higher air pressure toward places with lower air pressure, just as air flows out of a balloon or a tire when it gets a chance. The pressure force is directed 90° *across* an isobar, from the side with higher atmospheric pressure to the side with lower atmospheric pressure.

Air will be pushed away from high-pressure centers, like water flowing away from the top of a hill. Likewise, air will be pushed toward low-pressure centers, like water in a sink flowing lower toward the drain.

The pressure force always pushes from higher pressure to lower pressure—this is the direction in which it *pushes*. Though air is directed from high pressure to low pressure, in the atmosphere the air does not move the same way it is pushed.

How do isobars show the direction that atmospheric pressure pushes the air to flow? _____

Answer: Atmospheric pressure pushes air to move perpendicular across isobars from areas of higher pressure to areas of lower pressure.

Coriolis Force

Air does not simply move in the direction pressure pushes it because Earth is rotating on its axis. Because the planet spins, all things that move *across* the surface of Earth experience something known as the Coriolis effect. You can think of the spinning spherical Earth as the three-dimensionalization of a two-dimensional spinning disk like a merry-go-round or a carousel. If you imagine the path that a rolling ball would take across a spinning merry-go-round, then you're on the right track for conceptualizing how Coriolis affects things moving across the spinning Earth.

A thorough explanation of the Coriolis effect requires a lot of physics (which is given in appendix 2), but it is not necessary to master the calculations if, for now, you can simply accept the conclusions. Let's keep Coriolis as simple as we can; this is what you need to know:

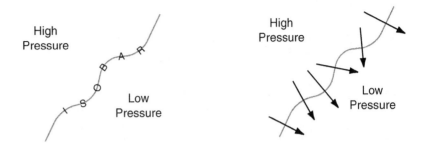

Figure 4.2. *Left:* Air pressure is identical all along an isobar, and the atmosphere will have higher air pressure on one side of an isobar than it does on the other side. *Right:* The pressure *force* will push 90° across the isobar—from the high-pressure side to the low-pressure side.

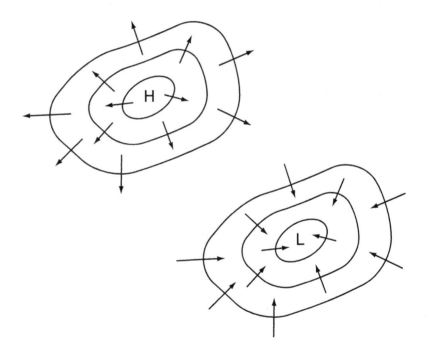

Figure 4.3. Pressure force always pushes from areas of higher air pressure to areas with lower air pressure. *Top:* Pressure pushes outward, across the isobars, away from a high-pressure center. *Bottom:* Pressure pushes inward, across the isobars, toward a low-pressure center.

- In the Northern Hemisphere, Coriolis will make moving things turn to the right.

- In the Southern Hemisphere, Coriolis will make moving things turn to the left.

- The faster that gases (e.g., the atmosphere) or liquids (e.g., ocean currents) move, the stronger the Coriolis turning force is.

- Things that go up or down are not moving across the surface of Earth, so the Coriolis force is not in effect.

The explanations provided in this chapter for air circulation around high- and low-pressure centers were developed for conditions in the Northern Hemisphere, where the Coriolis force makes an object turn to its right. To apply the explanations to the Southern Hemisphere, substitute "to the left" for "to the right" and be aware that the consequent clock-respective descriptions and accompanying illustrations would need to be reversed.

If a textbook tries to explain the Coriolis concept with a diagram of a cannonball or a missile flying over a spinning Earth, it is very likely that the explanation is not correct. The last section of appendix 2 has a non-technical discussion of that particular example.

The Coriolis will make moving things in the Northern Hemisphere turn to the _____ of the path they are heading, and in the Southern Hemisphere moving things turn to the _____.

Answer: right; left

High-Pressure Circulation

The Coriolis force always is directed perpendicular to an object's direction of movement. Once Coriolis is added, the wind can no longer cross isobars at an angle of 90°. The pressure force pushes the air that way, but that is not the way the air will end up moving. Once the pressure force sets the air in motion, Coriolis will kick in, and the Coriolis

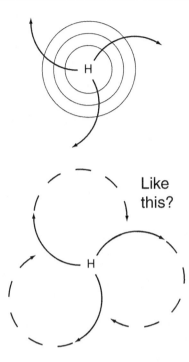

Figure 4.4. *Top:* At a high-pressure center, the pressure force starts the air moving from higher pressure to lower pressure, and then Coriolis will immediately kick in after the onset of movement. The Coriolis force is directed perpendicular to the direction of movement and will cause moving air to turn to the right (Northern Hemisphere) of its direction of movement. What is going to happen? Will the air go like this (*bottom*) if it has to keep turning to the right?

force will push the air to the right of its direction of movement. Coriolis always will try to turn a moving object; it is never satisfied. So air will not move in a straight line away from high pressure. The air will be pushed toward lower pressure and also turned to the right.

Even though Coriolis always is compelling a turn to the right, the object is prevented from circling all the way back because air cannot flow from low pressure to high pressure. *Air can never flow toward higher pressure.* Air always will be pushed away from high-pressure areas toward lower-pressure areas. If the air cannot flow in a straight line because Coriolis requires it to always be turning to the right, and if it cannot circle back from a lower-pressure area to a higher-pressure area, there must be an accommodation between the pressure force and the Coriolis force.

So, pressure pushes the air out and away; Coriolis turns it to the right. The wind cannot spiral back toward higher pressure. However, the wind cannot spiral outward either, because this would involve a turn to the left. The air cannot head in and it cannot veer out, but it *is* being pushed by air pressure differences, so it must move; the only resolution is to travel in a path that keeps it aligned with the isobars. Air will flow *around* a high-pressure system. Air rotates clockwise around a high-

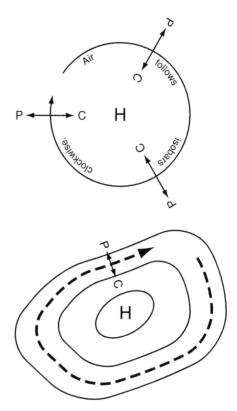

Figure 4.5. *Top:* The pressure force (P) is always directed away from higher pressure. The Coriolis force (C) is always directed to the right (Northern Hemisphere) of the direction of movement. This causes air to rotate clockwise, following the isobars, around a Northern Hemisphere high-pressure system. *Bottom:* The way to balance the high-to-low force and a turn to the right is for air to follow the isobars clockwise around the pressure center (Northern Hemisphere). The air will not cross the isobars because to do that, it must either flow toward higher pressure or turn to the left, neither of which is possible.

pressure center. In this way it can be continuously pushed by the pressure force and turned by Coriolis.

In what direction does air circulate around a high-pressure system in the Northern Hemisphere? _____

Answer: clockwise

Low-Pressure Circulation

In a low-pressure system, circulation is opposite from that found in a high-pressure system. The pressure force wants to drive air in toward the low-pressure center, but the air does not move in a straight line. The air is being pushed away toward lower pressure and also is turning to the right. The turn to the right is prevented from circling around because air cannot flow "upstream" to an area with higher air pressure. *Air can*

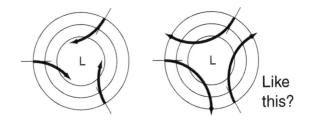

Like this?

Figure 4.6. *Left:* At a low-pressure center, if the pressure force initially has air moving from higher pressure to lower pressure, then Coriolis will affect the direction of movement. The Coriolis force is directed perpendicular to the direction of movement and will cause moving air to turn to the right (Northern Hemisphere) of its direction of movement. What is going to happen? Will the air go like this *(right)* if it has to keep turning to the right?

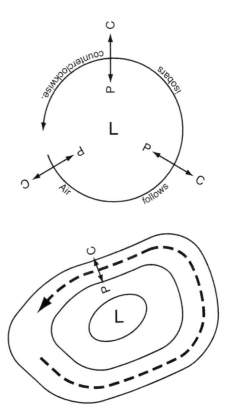

Figure 4.7. *Top:* The pressure force (P) is always directed from higher pressure toward lower pressure. The Coriolis force (C) is always directed to the right (Northern Hemisphere) of the direction of movement. This causes air to rotate counterclockwise, following the isobars, around a Northern Hemisphere low-pressure system. *Bottom:* The way air can balance the high-to-low force and a turn to the right is to follow the isobars counterclockwise around the pressure center (in the Northern Hemisphere). The air will not cross the isobars because to do that, it must either flow toward higher pressure or turn to the left, neither of which is possible.

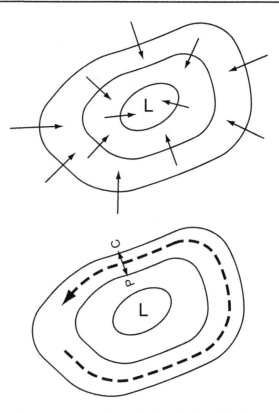

Figure 4.8. *Top:* Pressure pushes the air directly from higher-pressure areas to lower-pressure areas. *Bottom:* The Coriolis force will act to cause the air to circle counterclockwise around Northern Hemisphere low-pressure centers. Although it clearly appears that the air is curving to the left, in geographic terms Coriolis has the wind moving to the right of the direction in which the pressure wants it to go.

never flow toward high pressure. Air always will be pushed from high-pressure areas toward lower-pressure areas.

The pressure force wants to push the air straight across the isobars, but Coriolis will immediately force the air to turn right. If the air can't flow straight, and it also can't flow back from lower to higher pressure, then another compromise must be struck between the pressure and Coriolis forces.

So pressure pushes the air in toward the low-pressure center; Coriolis turns it to the right. The wind cannot spiral out toward higher pressure, but the wind cannot spiral in either, because this would require a turn to the left. The air cannot head inward and it cannot veer outward, but it *is* being pushed by air pressure differences, so it must move: as with high-pressure situations, the only resolution is to travel in a path that keeps it aligned with the isobars. An accommodation is reached by having air move counterclockwise around a low-pressure center.

Even though the air is depicted as turning to the left in Figure 4.7, it is actually being turned to the right by Coriolis. The air's path is caused by a pressure force directed toward low pressure *and* a Coriolis force that *is* turning the air to the right of the direction in which the pressure force wants to push it.

In the Northern Hemisphere, the wind balances the pressure force and the Coriolis force by rotating _____ around a low-pressure system.

   Answer: counterclockwise

Friction and Coriolis

There is an additional complication to circulation around high- and low-pressure systems. When the air that is rotating around a high- or low-pressure center is near the ground, it is influenced by the surface of Earth. There are buildings, trees, hills, and valleys, all of which produce friction. Friction causes resistance that slows down the speed of the moving air, which means that the Coriolis force is weaker (slower speeds = weaker Coriolis). So ground friction will cause the air near the surface to follow a slightly different path than the higher-level air that travels without ground friction.

Why is the Coriolis force weaker near the ground than higher up in the atmosphere? _____

   Answer: Friction with the ground slows the speed of the air. The slower the air moves, the weaker the Coriolis force becomes.

High-Pressure Circulation with Ground Friction

Initially air moves from higher-pressure places to lower-pressure places. The Coriolis force causes moving things to turn to the right. This will cause a clockwise rotating flow around high pressure.

Ground friction will affect the balance between the pressure and

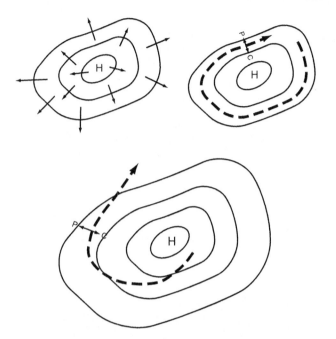

Figure 4.9. *Top, left:* First, air is pushed from higher-pressure areas to areas with lower air pressure. *Top, right:* Second, this pressure and Coriolis combine to make winds follow clockwise paths around the isobars encircling a high-pressure center (Northern Hemisphere). *Bottom:* When the speed of the air is slowed by ground friction, Coriolis still will make it turn to the right (Northern Hemisphere), but the turning influence will not be as strong. If the pressure force (P) is greater than the Coriolis force (C), then pressure will be able to overcome the turning effects of Coriolis and push the air out, away from a high-pressure center.

Coriolis forces (the balance is what creates the circular flow). When friction reduces the speed of the air near the ground, it reduces the amount of the Coriolis force generated. The force due to the *pressure* is unchanged, but the force due to the *Coriolis effect* is weaker.

With the same pressure force but a weaker Coriolis force, the forces will not balance each other out, and near the surface, where friction interferes with the clean flow of the wind, air will spiral *clockwise and out,* away from high–pressure centers.

What path does air near the ground follow in a Northern Hemisphere high-pressure system? _____

Answer: Air will spiral clockwise and out, away from the high-pressure center.

Low-Pressure Circulation with Ground Friction

A similar reverse pattern happens for air near the surface in low-pressure systems. Initially pressure drives air from high-pressure areas toward the low-pressure center. Coriolis will balance the pressure force to cause a counterclockwise rotating flow.

Ground friction will affect the balance between the pressure and Coriolis forces. When friction reduces the speed of the air near the ground, it reduces the amount of the Coriolis force generated. As before, the force due to the pressure is unchanged, but the force due to the Coriolis effect is weaker. The two forces will not balance each other out to cause a rotating flow of air, as they do when there is no

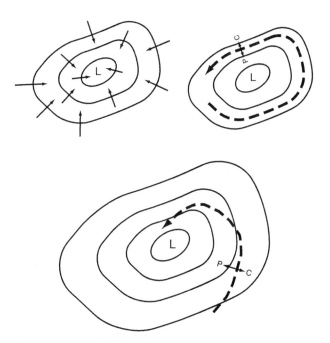

Figure 4.10. *Top, left:* First, air is pushed from higher-pressure areas to areas with lower air pressure. *Top, right:* Second, this pressure and Coriolis combine to make winds follow counterclockwise paths around the isobars encircling a low-pressure center (Northern Hemisphere). *Bottom:* When the speed of the air is slowed by ground friction, Coriolis still will make it turn to the right (Northern Hemisphere), but the turning influence will not be as strong. If the pressure force (P) is greater than the Coriolis force (C), then pressure will be able to overcome the turning effects of Coriolis and push the air in, toward a low-pressure center.

This visible-spectrum satellite image shows Hurricane Floyd (top left) on September 15, 1999, shortly before it made landfall in North Carolina; Hurricane Gert appears at lower right. Note the counterclockwise and inward spiral of the clouds in these two Northern Hemisphere, low-pressure storms. (Satellite image courtesy of National Climatic Data Center.)

friction. Near the surface, where friction interferes with the clean flow of the wind, air will spiral *counterclockwise and in* toward low-pressure centers.

What path does air near the ground follow in a Northern Hemisphere low-pressure system? _____

Answer: Air will spiral counterclockwise and inward, toward the low-pressure center.

Pressure Systems and Vertical Air Movement

An important concept that is associated with circulation around high- and low-pressure centers is that the circulation will cause air to rise or descend at the pressure center. When ground-level air moves away from a high-pressure zone, that does not create a zone without any air; no vacuum is created when air moves away from a high-pressure center. The air that moves away has to be replaced by air from somewhere else. Since surface air is moving away in all directions, air has to come from

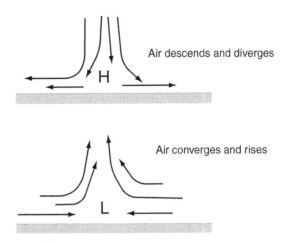

Figure 4.11. *Top:* High pressure is characterized by diverging air at the surface and replacement air descending from higher in the atmosphere. *Bottom:* Low pressure is characterized by converging air at the surface and air rising up into the atmosphere.

higher up in the atmosphere to replace the ground-level air that is moving out.

Similarly, when ground-level air moves toward a low-pressure center, that does not create a zone with an accumulation of excess air. The air that moves in has to go somewhere else. Since ground-level air is

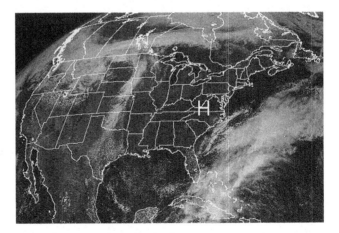

This visible-spectrum satellite image shows how high pressure centered near Virginia brings clear skies to the eastern United States on the afternoon of May 22, 2002. (Satellite image courtesy of National Climatic Data Center.)

coming from all directions, air has to rise up from the surface to relieve the buildup that would otherwise occur.

When surface air spirals in toward a low-pressure center, how is the accumulation of atmospheric material accommodated? _____

Answer: The surface buildup is relieved when air rises upward at a low-pressure center.

SELF-TEST

1. The pressure force pushes air _____.
   a. parallel to the isobars
   b. perpendicular, straight across the isobars
   c. at a 30° angle to the isobars
   d. from the low-pressure side to the high-pressure side of an isobar

2. Objects that move across the surface of Earth are subject to a Coriolis turning force that is caused by _____.
   a. friction
   b. the value of the atmospheric pressure
   c. rotation of Earth on its axis
   d. relative humidity

3. In the Northern Hemisphere, surface winds in a low-pressure system spiral _____.
   a. clockwise *and* inward
   b. clockwise *and* outward
   c. counterclockwise *and* inward
   d. counterclockwise *and* outward

4. Isobars are the lines on a map that are used to connect places that have the same _____.

5. Air sinks and diverges at a _____ pressure center.

6. As an object's speed drops, the strength of the Coriolis turning force that is pushing on that object also decreases. (True or False)

7. In the Northern Hemisphere, Coriolis force makes moving things turn to the right of their direction of travel. (True or False)

8. What causes the pressure and Coriolis forces to become unbalanced when air is moving near the ground?

9. Why does air rise at a low-pressure center?

10. Why would a high-pressure center be associated with light and variable surface winds?

ANSWERS

1. b   2. c   3. c   4. air pressure or atmospheric pressure

5. high-   6. True   7. True

8. As air moves near the ground, friction slows the airspeed, which reduces the Coriolis force. The pressure force is unaffected.

9. Air is converging from all directions toward a low-pressure center. The only way to relieve the surface buildup is for the air to rise.

10. Air is descending at a high-pressure center and then moving away in all directions. There is no dominant direction of horizontal air movement.

Links to Other Chapters

- Ascending air in low pressure is associated with precipitation (chapter 3).
- Parts of the general circulation of the atmosphere are established by the push of air from high pressure to low pressure and its turning by Coriolis (chapter 5).
- Air descending at high-pressure areas and ascending in low-pressure areas is an important part of the general circulation of the atmosphere (chapter 5).
- Extratropical cyclones are low-pressure storm systems (chapter 6).
- Tropical storms and hurricanes are low-pressure storm systems (chapter 6).
- Descending air in high-pressure systems is associated with clear, dry weather and deserts (chapter 7).
- Moving air is a geomorphic agent that can mobilize and transport sediment (chapter 14).
- Wind is the principal force in the generation of ocean waves (chapter 15).

5 General Circulation of the Atmosphere

Objectives

In this chapter you will learn that:

- The general circulation is caused by uneven heating of Earth and the Coriolis force generated by Earth's rotation on its axis.

- Near latitude 30°, air sinks from the upper parts of the troposphere in places called subtropical high pressure.

- In tropical areas, surface trade winds blow from east to west.

- Near the equator, air rises when trade winds from each hemisphere come together at the intertropical convergence zone.

- In midlatitude areas, surface westerlies blow from west to east.

- Near latitude 60°, air converging from surface westerlies and polar easterlies rises to form the subpolar low-pressure belt.

- The elements of the general circulation of the atmosphere move slightly north and south as the places with the greatest heating change from season to season.

Up, Down, Up, Down; East, West, East

Before discussing the origin and characteristics of the global system of air circulation let's set out the characteristics of general circulation. On a planetwide scale, there are seven elements of atmospheric circulation in each hemisphere that affect the surface of Earth. Four elements have air rising or sinking, while three of the elements are surface winds.

Air is rising or descending at (1) the intertropical convergence zone (ITCZ); (2) the subtropical high-pressure zone, or the subtropical high; (3) the subpolar low-pressure zone, or the subpolar low; and (4) the polar high-pressure zone, or the polar high. This is not so difficult to remember—think 0, 30, 60, 90—then think *up, down, up, down,* and you have it. It's the same in each hemisphere (they share the *up* at 0°).

1.	Intertropical convergence zone	0°	Up
2.	Subtropical high-pressure zone	30°	Down
3.	Subpolar low-pressure zone	60°	Up
4.	Polar high-pressure zone	90°	Down

In general, every 30°, starting at the equator, air goes *up* or *down.*

Now we'll add two things that you already know from chapter 4. When wind converges it must rise, and rising air is a feature of low pressure. When air descends, that causes high pressure and diverging winds. So *up, down, up, down* is analogous to *low, high, low, high* pressure.

The intertropical convergence zone is literally an area of converging air located between the Northern Hemisphere tropics and the Southern Hemisphere tropics. Converging air means low pressure, rising air: the ITCZ is low pressure around 0° latitude. The subtropical high-pressure zone is high pressure located subtropically—that is, a bit outside of the tropics. High-pressure, descending, diverging air is found around latitude 30°. The subpolar low is low pressure located subpolarly—that is, not quite in the polar latitudes. There is converging, rising air at around latitude 60°. The polar high is simply a zone of high pressure, with descending, diverging air at around latitude 90°.

Between the four *up, down* elements in each hemisphere are three places where the wind moves across the surface of Earth. Remember

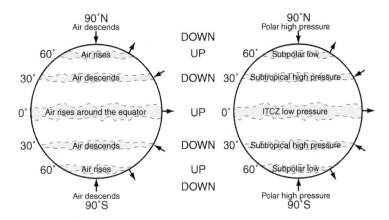

Figure 5.1. In the general circulation of the atmosphere, air goes up or down along 0°, 30°, 60°, 90° latitude in each hemisphere. The 30° up-down pattern corresponds to areas of high atmospheric pressure and low atmospheric pressure.

that winds are named for the direction they come *from*; think *east, west, east* and you have it.

    0°–30°   Easterlies (between up at 0° and down at 30°)

    30°–60°   Westerlies (between down at 30° and up at 60°)

    60°–90°   Easterlies (between up at 60° and down at 90°)

The easterlies between 0° and 30° are called trade winds: *tropical trades*. The midlatitude winds are creatively called the westerlies. The other easterly wind system on Earth is called the polar easterlies, to distinguish it from the tropical easterlies (the trade winds).

When you put the *up, downs* and the *east, wests* together, you have almost the entire model (remembering it's the same in each hemisphere). The characteristics of the general circulation of the atmosphere in each hemisphere can be summarized as:

0°, 30°, 60°, 90°: *up, down, up, down: low, high, low, high:*

and in-between winds blow *from*

east, west, east: tropical, middle, high latitudes

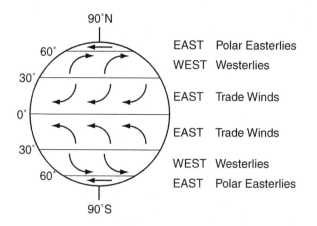

Figure 5.2. In the general circulation of the atmosphere, surface winds on Earth flow from the east in the tropics, from the west in the midlatitudes, and from the east in the high latitudes.

What does the general circulation of the atmosphere predict for air movement at the equator, 30°, 60°, and the pole? _____

Answer: In each hemisphere, the model has rising air at the equator, sinking air at 30°, rising air at 60°, and descending air at the pole (up, down, up, down).

General Circulation Model

The general circulation model explains the way that air circulates on Earth. There are some seasonal variations, but this model accounts for nearly everything major. This is an observable, real–world pattern. You do not need to know any of the reasons why circulation is like this to verify the direction that prevailing winds come from, or to verify that there are global belts of higher or lower pressure at certain latitudes. The pattern exists.

First there are some clarifications and definitions that are important for a discussion of air movement:

• "Up": away from the center of Earth.

• "Down": toward the center of Earth.

- "Turns to the right": as perceived by the item in motion. (For example, if, while you are driving, another car approaches head-on and its driver makes a right turn, you will observe the car moving to the left—but as perceived by the driver of the other car, it is turning to the right.)

- "Turns to the left": the item in motion perceives that it is turning left.

- Winds are named for the direction they blow *from*. A "west wind" blows from the west toward the east. The cold "north wind" blows from the north to the south.

The general circulation model is a simplification of the way that the global wind system works. The explanation is substantially correct, but two simplifications are needed to create an imaginary Earth for the model.

Simplification 1: The imaginary Earth's axis is not tilted. This means that the Sun will be most directly overhead at the equator every day, and there are no changes of season.

Simplification 2: There are no land and water differences to the surface of the imaginary Earth; the surface is the same everywhere.

We'll revisit the assumptions after we go through the model.

If air is said to be rising, or moving up, what does that literally mean?

_____

Answer: The air is moving away from the center of the spherical Earth.

Heating and the Equator

An explanation of the general circulation of the atmosphere could begin at any location, but it is intuitive to start at the equator. In the model, incoming solar radiation will be greater at the equator than at any other place, so the equatorial region will be hotter than other places. When the equatorial region gets hot, that warms the atmosphere, and the hot air rises up (moves away from the center of Earth).

As the air reaches the top of the troposphere, its rise is curtailed by that "ceiling." The troposphere remains separate from the stratosphere

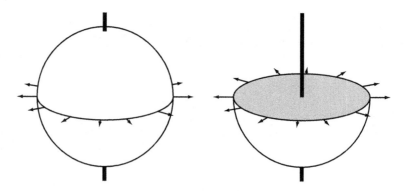

Figure 5.3. The Sun heats Earth, and the hotter air at the equator rises up (i.e., moves away from the center of Earth).

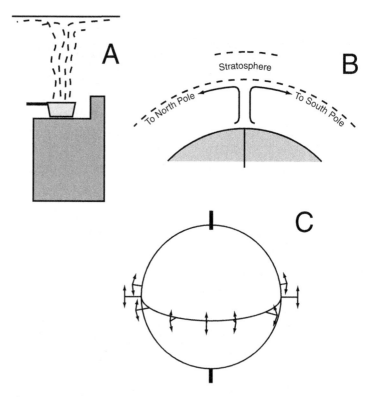

Figure 5.4. *A:* Rising steam above a hot stove is deflected to move horizontally by a kitchen ceiling. *B* and *C:* Rising hot air at the equator hits the troposphere ceiling and is deflected toward the North and South Poles.

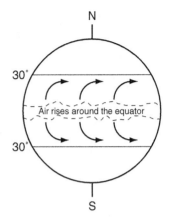

Figure 5.5. High-level air moving from the equator toward the poles is turned to the right (Northern Hemisphere) and to the left (Southern Hemisphere) of its path by Coriolis. Air ends up converging high in the troposphere, around 30°N and 30°S.

layer above it. Therefore the rising air will be directed toward the North or South Pole. Once deflected toward either pole, the air travels parallel to the surface of Earth (although high up in the troposphere); things that move parallel to the surface of Earth are affected by the Coriolis force. In the Northern Hemisphere, Coriolis will make the moving air turn to the right. In the Southern Hemisphere, Coriolis turns moving air to the left.

Why does rising air move away from the equator toward the poles?

Answer: As the rising air reaches the top of the troposphere, it is deflected away, and the only directions in which it is free to move are north or south.

Descending Air at 30°: Subtropical High Pressure

The air moving at a high altitude away from the equatorial zone in the Northern Hemisphere heads north toward the pole and turns to the right. At around latitude 30° the air "piles up." Coriolis has turned the air as it departed the equatorial area, and it is now converging around 30°. Earth also is smaller around at latitude 30°, so there is not enough room for all the air from the equator.

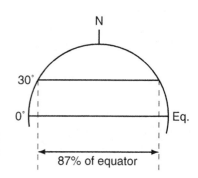

Figure 5.6. The diameter of Earth at 30° latitude is 87 percent as big as it is at the equator.

If there is air continuously accumulating and not enough room for it, the only place it can go is down. The air cannot go back toward the equator because it would have to go against the flow of air that is always coming *from* the equator. It cannot head to the pole because it would have to turn against the direction of the Coriolis force. It cannot just go east and circle Earth because there is not enough room at that latitude for all the air that is constantly coming from the equator. Because it will not cross into the stratosphere the air must descend back to the surface; the only release is down.

Descending air around latitude 30° creates high pressure (descending, diverging air). When the sinking air gets to the surface at that lat–

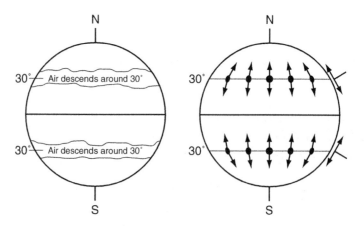

Figure 5.7. *Left:* There is a belt of descending air around Earth at 30°N and 30°S. This is the subtropical high-pressure region. *Right:* When the descended air reaches the surface in the subtropical high, the air will be pushed north or south, away from the high pressure.

itude, it must move outward, either toward the pole or toward the equator.

The air that is pushed away from the subtropical high is once again moving across the surface of Earth and will now be affected by the Coriolis force, which will turn the wind:

- to the right in the Northern Hemisphere

- to the left in the Southern Hemisphere

After air descends at latitude 30°, where does it move next?

_____

Answer: Air is pushed toward the pole and toward the equator.

Trade Winds

Let's first take the case of air that is pushed by the subtropical high pressure toward the equator (in the upcoming "Westerlies" section we'll get back to the winds that are pushed poleward). Surface winds are pushed out toward the equator by the subtropical high pressure. Once the pressure force gets it moving, Coriolis will make it turn. In each hemisphere, the wind is flowing from the area around 30° toward the equator. In each hemisphere, the wind will be turned to flow toward the west by the Coriolis force. These easterly winds (from east to west) in tropical regions are called trade winds.

In the Northern Hemisphere, air is pushed southward from the subtropical high and turns to the right. The wind approaches the equator from the northeast and is called the northeast trade winds. In the Southern Hemisphere, air is pushed northward from the subtropical high, and then Coriolis turns it to the left. This movement to the north and west is why these winds are called southeast trade winds (from the southeast). There is now converging air at the equator. Northeast surface winds are coming from the Northern Hemisphere and southeast surface winds are coming from the Southern Hemisphere, and they are meeting at the equator.

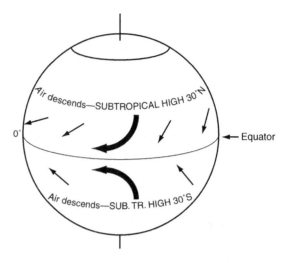

Figure 5.8. Air pushed away from the subtropical high pressure flows toward the equator and will be turned by Coriolis to become the easterly trade winds.

Why do the trade winds come from an easterly direction in each hemisphere? _____

Answer: In the Northern Hemisphere, winds pushed toward the equator are turned to the right (air moving south is turned to travel west), and in the Southern Hemisphere, winds pushed to the north are turned to the left, to head westward. Air moving westward comes from the east.

Intertropical Convergence Zone

The area where the trade winds from each hemisphere meet is called the intertropical convergence zone. Intertropical means "between the tropics"—the place where wind coming from the Northern Hemisphere tropical latitudes meets with wind coming from the Southern Hemisphere tropical latitudes. When air converges, it has to go somewhere; in this case, because it is already at the surface, it has to go up.

When air at the equator converges and rises, it completes the cycle that we began with—ascending air at the equator. The equator is a low-pressure region—air is converging and rising. The rise also is supported by the large amount of heating and convection caused by direct insola-

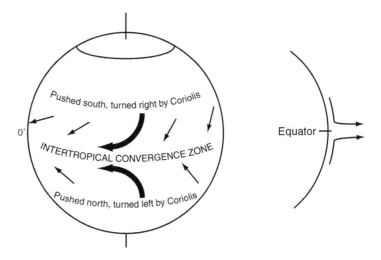

Figure 5.9. *Left:* Northeast trade winds from the Northern Hemisphere and southeast trade winds from the Southern Hemisphere will converge near the equator at the intertropical convergence zone (ITCZ). *Right:* Air converging at the ITCZ has to rise to relieve the accumulation of air coming from each hemisphere.

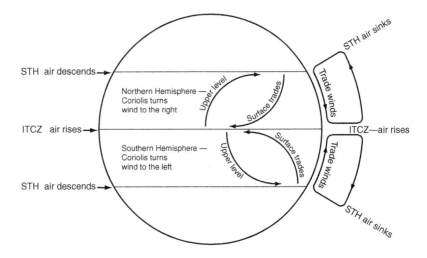

Figure 5.10. The movement of air in an up-poleward-down-return cycle between the intertropical convergence zone and the subtropical high in tropical regions is called Hadley circulation.

tion. The circulation of air (1) up at 0°, the equator, (2) poleward at high altitude, (3) down in the subtropical high-pressure zone at around 30°, and (4) across the surface back to the equator as the trade winds, is called a Hadley cell.

What is converging at the intertropical convergence zone?

_____

Answer: Trade winds from the Northern and Southern Hemispheres

Westerlies

What happens on the pole side of the subtropical high-pressure zone? After descending air reaches the surface in the subtropical high-pressure belt (approximately 30°), some is pushed by the high pressure toward that hemisphere's pole. Once the pressure starts the air moving across the surface of Earth, Coriolis will immediately make it turn (to the right in the Northern Hemisphere, to the left in the Southern Hemisphere). In each hemisphere, the surface wind that was initially heading to the pole will turn to flow from west to east; these winds are called westerlies.

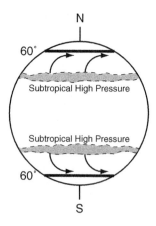

Figure 5.11. In each hemisphere, wind pushed toward the poles from the subtropical high-pressure belts is turned by Coriolis to flow from west to east.

For a midlatitude city such as Lincoln, Nebraska (41°N), what would be the prevailing global wind system? In what direction is the air moving?

_____

Answer: Air moves from west to east in the prevailing westerlies.

High Latitudes

The midlatitude surface westerlies begin to pile up around 60°. Converging surface air is filling into a smaller place (the same way that upper air in the Hadley cell piles up around 30°). When air at the surface needs more room, it can go only one place: up. It rises. The air cannot head back toward lower latitudes because it would have to flow against the persistent influence of subtropical high pressure. The air cannot head toward the pole because it would have to turn against the Coriolis force. The air cannot just perpetually build into 60°. The only escape is for the air to rise.

When the westerlies from the subtropical high-pressure zone converge and rise at 60° latitude, that means air moves up, away from the center of Earth. The converging, rising air creates a low-pressure zone around Earth at 60° latitude in each hemisphere. This region is called the subpolar low-pressure zone. As the air rises there and hits the troposphere ceiling, it is diverted to flow toward the pole and toward the equator.

First we will consider the upper-level air moving away from the subpolar low toward the pole. We will examine the upper-level air heading toward the equator later in this section.

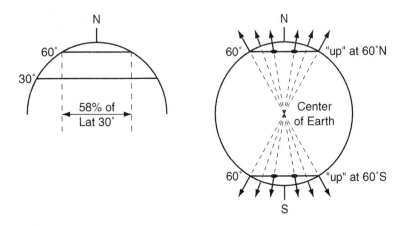

Figure 5.12. *Left:* The distance around Earth is much smaller at 60° latitude than it is at 30°. *Right:* The surface westerlies converge at around 60° latitude, and the air rises (moves away from Earth's center) to relieve the buildup.

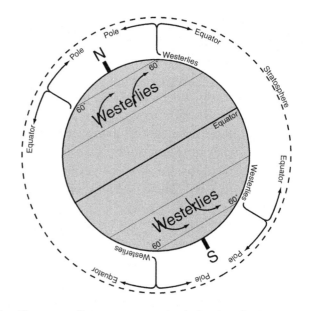

Figure 5.13. The westerlies converge at 60° latitude, which causes the air to rise. When the rising air in the subpolar low at 60° hits the troposphere ceiling, it is diverted at high altitude and moves toward the pole and the equator.

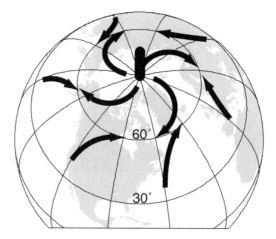

Figure 5.14. After air descends in high pressure at the pole, it can only move away, toward the equator. Coriolis will make these surface winds in the Northern Hemisphere turn to the right to become the polar easterlies. This surface convergence of air coming from the polar high will reinforce the convergence and rising that the surface westerlies set up in the subpolar low-pressure zone around 60°.

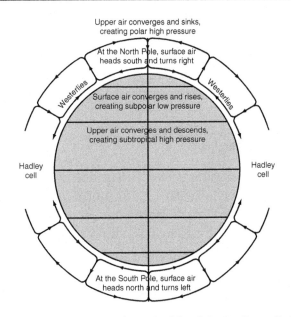

Figure 5.15. The air circulation on the pole side of the Hadley cell affects the surface with four elements of the general circulation: (1) the midlatitude westerlies, (2) the subpolar low, (3) the polar high, and (4) the polar easterlies. Upper-level circulation reinforces the descending air of the subtropical high-pressure area.

The upper-level air that is directed toward the pole will be arriving there from every direction. The air at the pole literally has nowhere to go but down—a descent to the surface is the only release. After the descending air reaches the ground at the pole, there is only one way for the air to go—toward the equator. Descending air creates high pressure and diverging winds, the polar high.

In each hemisphere, after the polar high pushes the surface air back toward the equator, Coriolis will make the air turn to head from east to west. These surface winds are called polar easterlies. These polar easterlies will converge with the midlatitude westerlies around 60°.

Now let's consider the other upper-level stream that rises at 60° in the subpolar low. It heads toward the equator, is turned by Coriolis, and builds at high altitude around latitude 30°. This is the same place where upper-level air moving away from the equator is accumulating. Thus there is a big pileup of air in the upper troposphere at 30°. Upper air moving away from the subpolar low converges with upper circulation in the Hadley cell; this collision reinforces the downward flow of air at 30° that creates subtropical high pressure.

In the general circulation model, what happens at Earth's surface at latitude 60°? _____

Answer: Air converging from the midlatitude westerlies and the polar easterlies creates the subpolar low-pressure zone, where air rises.

Seasonal Changes

This general circulation model explains how global winds work on the imaginary Earth. But remember the two simplifications from "General Circulation of the Model" earlier in this chapter? On the real Earth things are basically the same, but they are a little more complicated for two reasons:

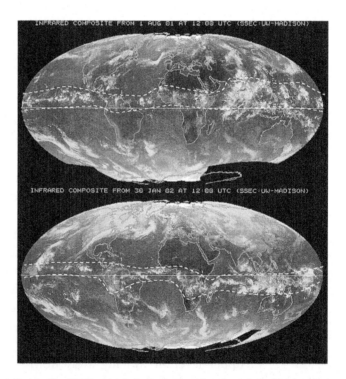

In August *(top)*, the ITCZ, which is indicated by the global belt of cloudiness in tropical latitudes, is in the hotter Northern Hemisphere. In January *(bottom)*, the ITCZ has followed the Sun into the Southern Hemisphere. The ITCZ will move slightly from day to day, but this pair of images is representative of the height of summer for each hemisphere. (Base images courtesy of Space Science and Engineering Center, University of Wisconsin-Madison.)

- The equator is not always the place that receives the most insolation. Earth's axis is tilted so the latitude receiving the maximum insolation will change from day to day.

- Land and water heat at different rates (chapter 2); thus the effects of insolation vary even at the same latitude. So the intertropical convergence zone (ITCZ) will actually be found at different latitudes in different parts of the world.

If the place with greatest heating moves from day to day, that means the elements of the general circulation will move with it. The general circulation elements are "attracted" by the Sun but anchored at each pole. They will stretch their locations north and south to match the location of maximum heating. The equator isn't always the hottest place. The place with the most direct insolation moves. From March to September, the most direct insolation is in the Northern Hemisphere. From September to March, the most direct insolation is in the Southern Hemisphere. The seasonal movement of the most intense insolation will move the location of the ITCZ, which will affect the location of all the other elements of the general circulation. The yearly north–south sequence of approach and recession of the general circulation elements can have a dramatic impact on place's climate.

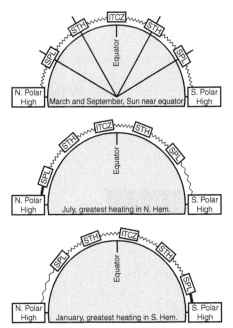

Figure 5.16. *Top:* Around March and September, when the strongest insolation is near the equator, the general circulation elements are about where the model predicts them to be. *Center:* In July, when the maximum insolation is well into the Northern Hemisphere, the general circulation elements will migrate northward to match the strong sunlight. *Bottom:* In January, when the maximum insolation is well into the Southern Hemisphere, the general circulation elements will move south to follow the Sun.

When the Sun is over the equator in the spring and fall, the elements of global circulation are about where the general model predicts them to be.

In the Northern Hemisphere summer, the general circulation elements stretch their locations to the north.

- The ITCZ follows the Sun and moves into the Northern Hemisphere.

- The trade winds shift a bit north in response to the shift of the ITCZ.

- The subtropical high moves a bit north and will affect more of the continental United States.

- The westerlies move a bit north.

- The subpolar low moves a bit north, which spares the United States many storms.

In the Northern Hemisphere winter, the general circulation elements follow the Sun southward.

- The ITCZ moves into the Southern Hemisphere, with the zone of greatest heating.

- The trade winds shift south in response to the ITCZ movement.

- The subtropical high moves south, away from the United States.

- The westerlies and the subpolar low move south and can cause winter storms in the United States.

What two factors cause the location of the general circulation elements to move over the course of a year? _____

Answer: (1) seasonal changes to the place with maximum insolation; (2) differences in the rate of heating between oceans and land

SELF-TEST

1. In the Southern Hemisphere, what best describes the flow direction of the trade winds?
   a. From west to east
   b. from southeast to northwest
   c. up
   d. from north to south

2. On the pole side of the subtropical high-pressure zone in the Northern Hemisphere, surface air is moving _____.
   a. to the north and east   c. up
   b. south   d. to the west

3. Which element of the general circulation of the atmosphere would most affect a city such as Lincoln, Nebraska (lat. 41°N)?
   a. polar high   c. ITCZ
   b. trade winds   d. westerlies

4. The cycling of air in the Hadley cell involves circulation in which air moves upward at the _____, and back down to the surface at the _____.

5. The prevailing winds in tropical areas are the _____.

6. The hottest temperatures on Earth are always at the equator. (True or False)

7. In the general circulation model, air is descending around 30°N and 30°S latitude. (True or False)

8. In the Northern Hemisphere summer, why might the ITCZ be found farther north over continents than over oceans?

9. What element of the general circulation of the atmosphere is the source for the surface westerlies?

10. If low pressure is associated with cloudy, rainy skies, then over the course of a year what might cause there to be rainy and dry seasons near the equator?

ANSWERS

1. b   2. a   3. d   4. ITCZ; subtropical high-pressure zone

5. trade winds   6. False   7. True

8. Because landmasses can heat up more efficiently than water, when the Sun is overhead far from the equator, the landmasses will be much warmer than the adjacent ocean.

9. The westerlies are produced when air is pushed toward the pole by the subtropical high; then the air is turned by the Coriolis force to flow to the east.

10. The ITCZ sweeps back and forth, following the Sun and the zone of greatest heating. When it comes near, it brings cloudy, rainy, thunderstorm weather. When it is farther away, skies are clearer.

Links to Other Chapters

- Uneven heating of Earth is caused by the tilt of its axis and the yearly revolution of Earth around the Sun (chapters 1 and 2).
- Landmasses heat up and cool down much easier than oceans (chapters 2 and 7).
- The troposphere and the stratosphere do not have much mixing; they stay as separate layers of the atmosphere (chapter 3).
- Rising air in low-pressure zones will promote precipitation (chapters 3 and 4).
- Air is pushed from high-pressure areas to low-pressure areas (chapter 4).
- High pressure is associated with sinking air and clear skies (chapter 4).
- Coriolis makes moving things turn right in the Northern Hemisphere and left in the Southern Hemisphere (chapter 4).
- Midlatitude and polar air masses will be pushed from place to place by prevailing winds (notably by the westerlies in the lower forty-eight United States) (chapter 6).
- North–south movement of the elements of the general circulation will affect climate and weather (chapter 7).

6 Air Masses and Storms

Objectives

In this chapter you will learn that:

- Air masses are large, coherent bodies of air that acquire the temperature and moisture properties of their source regions.

- Air masses meet each other at fronts.

- A cold front is where a cold air mass is advancing and lifting warm air off the ground.

- A warm front is where a warm air mass is advancing into space occupied by a cold air mass.

- Fronts tend to produce precipitation because lifted warm air cools and loses its ability to hold water vapor.

- Extratropical cyclones are midlatitude, low-pressure storm systems that form along fronts and can produce much rain or snow.

- Tropical storms and hurricanes are low-pressure storm systems.

Air Masses

Most of the day-to-day weather in the lower forty-eight United States is produced by the movement and interaction of air masses. An air mass is a large (hundreds to thousands of miles across) body of air that has similar, though not necessarily identical or uniform, temperature and moisture levels throughout.

The qualities of the atmosphere in an air mass are alike because air masses form over large, relatively uniform areas with only small terrain differences. Oceanic source regions will produce wet air, whereas continental source regions will have drier air. Air temperature depends on climatic conditions. For example, source regions closer to the equator will produce warmer air masses.

Generally, air will acquire the characteristics of its source region (hot, warm, cool, cold, wet, dry). When the air mass moves away from its source region, it brings the characteristics of the source region with it. As an air mass moves to a new place, it stays together as one unit, it does not break into pieces. This means that air masses will not mix with each other. Instead, as they meet, a boundary will form between the two air masses.

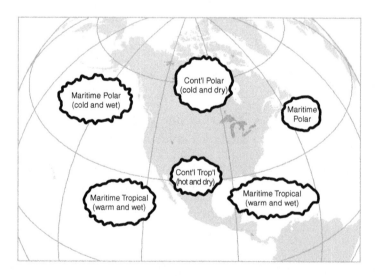

Figure 6.1. The continental United States is affected by air masses that form in source regions.

If an air mass forms over the northern Pacific Ocean near Alaska, what will its temperature and moisture qualities be? _____

Answer: The air mass will tend to be cold and wet.

Fronts

The boundary between two air masses is a "front." Just as in a war, the front is the line separating two antagonists that don't mix. There will be a defined boundary where they meet.

There are four kinds of fronts: cold, warm, stationary, and occluded. When referring to air temperature, "warm" and "cold" are relative terms that mean different things to different people. A warm front refers to warmer air that is moving forward—*warmer* than the air it is displacing (not necessarily warm in temperature). Likewise for cold fronts—it just means the new air will be *colder* than the prior air mass was.

Cold Front

A cold air mass pushes into a place that has warmer air. Cold fronts are indicated on maps with triangles that point in the direction the cold air is moving. On color maps they are blue.

Warm Front

A warm air mass moves into a place that currently harbors a cooler air mass. Warm fronts are mapped with semicircles that are shown leading the advance of the warmer air. On color maps they are red.

Figure 6.2. A front indicates the boundary between two distinct air masses.

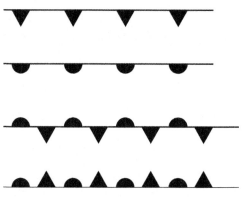

Figure 6.3. *From top:* Cold front, warm front, stationary front, occluded front.

Stationary Front

As the name suggests, a stationary front is a nonmoving boundary between warmer and colder air masses. The warm air is trying to move into the area occupied by the colder air mass. The cold air mass is trying to displace the warmer air. Each air mass is strong enough to resist the movement of the other, so the boundary remains in place. When mapped in color, stationary fronts have alternating blue and red sections.

Occluded Front

This is a special case that forms in extratropical cyclones. At the same place at the same time, warm air and cold air are very briefly moving in the same direction. Sometimes they are purple on color maps.

What kind of boundary would be found when one air mass is pushing into the space occupied by an air mass that has warmer temperatures?

――――――――――

Answer: cold front

Air Mass Interaction at Fronts

When an air mass pushes against another air mass, certain rules of engagement are followed.

- Air masses stay coherent; they don't split apart into smaller pieces.

- Warmer air will always end up on top of cooler air.

- If air is lifted, it will cool adiabatically, which can lead to clouds or precipitation.

Stationary Front

A stationary front is simple. Neither air mass is able to move forward into space occupied by the other, so the boundary is "stationary."

Cold Front

A cold front develops when a colder air mass begins to move into areas with warmer air. The cold air wedges into the warmer air and lifts it off the surface. As the warm air is lifted, it cools adiabatically and therefore loses its ability to hold water vapor. Water condenses to form clouds, which can lead to rain or snow. So after a cold front passes by:

- It will be colder since the cooler air mass will replace the warmer air mass.

- It will likely rain or snow, as the warmer air is forced upward.

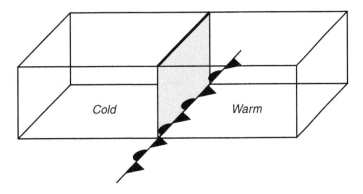

Figure 6.4. At a stationary front, each air mass resists the movement of the other, so the boundary between colder and warmer air masses stays put.

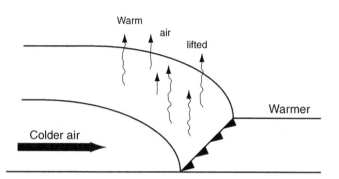

Figure 6.5. A cold front indicates that an advancing cold air mass is displacing warm air and wedging the warm air up, off the surface.

Warm Front

A warm front develops when a warmer air mass begins to move into areas with colder air. The overtaking warmer air rides up over the colder air. As the warmer air ascends, it cools adiabatically and therefore loses its ability to hold water vapor. Water condenses to form clouds, which can lead to rain or snow.

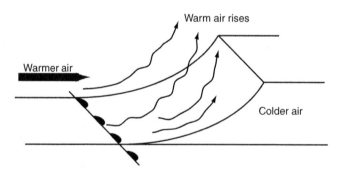

Figure 6.6. A warm front develops where a warmer air mass pushes out a colder air mass. Warmer air will ride up, over the colder air mass.

Why are fronts associated with rain or snow? _____

Answer: Warm air is being lifted off the surface, and as air rises, it cools and loses its ability to hold water vapor.

Extratropical Cyclones I

Extratropical cyclones are complex storm systems that begin when a low-pressure center develops on the boundary between two air masses. Extratropical means not in the tropics and refers to the midlatitudes. (A tropical cyclone is a "tropical storm" or a "hurricane.") An extratropical cyclone starts with a stationary front, then generates cold fronts, warm fronts, and eventually occluded fronts, which all revolve around a low-pressure system.

An extratropical cyclone begins along a stationary front between a warmer and a colder air mass. In North America, the colder air mass usually is north of the warmer air mass. The colder air will be trying to push south as the warmer air tries to move northward. Because of complicated upper atmosphere actions (whose explanation can safely be skipped), a low-pressure system can develop on the front. Because air rotates counterclockwise around a Northern Hemisphere low-pressure system, the colder air will start to push into the warmer air mass as the warmer air mass starts to move into the colder air zone. In an extratropical cyclone, the colder air moves faster than the warmer air, so the cold front begins to circle around the low-pressure center and catch up to the warm front.

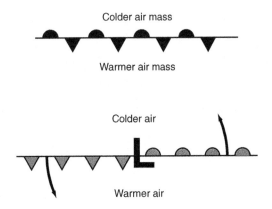

Figure 6.7. *Top:* The first step in the development of an extratropical cyclone is the formation of a stationary front between two air masses. *Bottom:* The extratropical cyclone forms when low pressure develops at the front and the two air masses begin to move counterclockwise (Northern Hemisphere) around the low pressure.

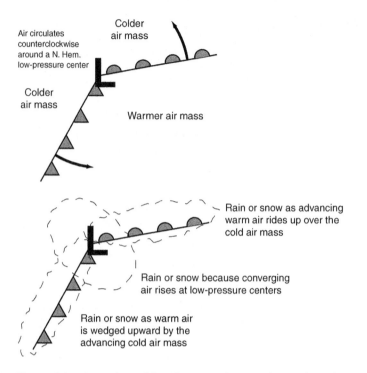

Figure 6.8. *Top:* The colder air mass advances faster than the warmer air mass, so it will circle around the low-pressure center faster than the warm air mass does. *Bottom:* Extratropical cyclones can produce a lot of rain or snow (depending on temperatures).

Extratropical cyclones can produce a lot of precipitation (rain or snow, depending on temperatures) because of the lifting at the warm front, at the cold front, and at the low-pressure center.

What is the initial condition that leads to the development of an extra-tropical cyclone? _____

Answer: a stationary front between cold and warm air masses

Extratropical Cyclones II

Eventually, the cold front at the lead of the cooler air mass begins to catch up with the warm front. The warmer air mass is squeezed into

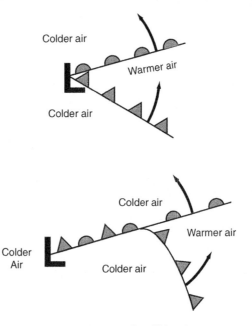

Figure 6.9. *Top:* As the cold air advances, it will begin to occupy more and more of the area surrounding the low-pressure center, and it will start to catch up to the advancing warm front. *Bottom:* An occluded front indicates that the advancing cold air mass has caught up to the advancing warm front and that the cold air mass has circled around and closed up with itself.

This infrared satellite image from 7:00 P.M. EST on January 27, 1998, shows warmer temperatures in dark grays and cold temperatures in light grays. Cloudiness is indicated by the lightest shades because the tops of clouds are colder than the surface of Earth is. An extratropical cyclone is centered near the North Carolina/South Carolina border and is evolving into a coastal storm called a northeaster. Clouds and rain are associated with the two fronts and with the low-pressure center; snow is falling in the colder sections of this storm. (Base satellite image courtesy of National Climatic Data Center.)

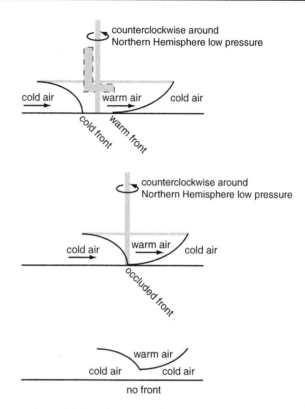

Figure 6.10. As viewed from the side: *Top:* The cold air mass circling around the low-pressure center begins to close in on the warm air mass, which is also rotating around the low. *Center:* An occluded front indicates that an advancing cold air mass has caught up to a warm front. *Bottom:* Once the warm air mass is completely wedged off the ground, and the cold air mass rejoins with itself, there are no more fronts. No air masses are moving into areas occupied by other air masses.

an increasingly smaller area between two parts of the same cold air mass. At the moment the advancing cold air closes with the rest of its air mass there is simultaneously advancing warm air and advancing cold air at the same place. This is the situation referred to as an occluded front.

When the advancing cold air closes with the rest of the colder air mass, this squeezes all of the warm air off of the ground. Occlusion shuts down the cyclone circulation because there is no longer a front between two unlike air masses. The cold air swings around the low-pressure center and is rejoined with colder air from the same colder air mass.

Why can the two cold sectors in an extratropical cyclone unite to lift the warm air off the surface? _____

Answer: The two cold sectors are part of the same cold air mass.

Tropical Storms and Hurricanes

Other kinds of low-pressure weathermakers are tropical storms and hurricanes. The only real difference between a tropical storm and a hurricane is intensity. A tropical storm is a very-low-pressure storm that starts in tropical regions. There are not significant differences in atmospheric characteristics within the tropics, so air mass interactions are not very important in those regions. Tropical air will be warm year-round, and due to its temperature, it can hold quite a bit of moisture, especially in marine areas. Tropical storms form and move without fronts. Like all low-pressure storms, in tropical storms:

- Surface winds spiral counterclockwise and inward toward the low-pressure center in the middle of the storm (in the Northern Hemisphere).

- Wind speeds are very high because there is very low pressure. Very low pressure means that there is a very strong pressure force, and when speeds are high, there is a very strong Coriolis force.

- Near the center, the converging air rises.

- Rising air cools and leads to intense rain because warm, maritime tropical air holds a tremendous amount of moisture.

Why do hurricanes produce so much rain? _____

Answer: Hurricanes form in warm air that can hold a lot of water vapor and in ocean areas where a lot of moisture is available. When that warm, moist air is lifted in the low-pressure center it can cool and become saturated; the water vapor condenses to become rain.

SELF-TEST

1. A hurricane is most likely to form in a _____ air mass.
   a. continental tropical (cT)   c. maritime tropical (mT)
   b. continental polar (cP)   d. maritime polar (mP)

2. An occluded front is formed in an extratropical cyclone when
   _____.

   a. a cold front catches up to a   c. a cold front evolves into a
      warm front   stationary front
   b. a warm front catches up to a   d. a stationary front evolves
      cold front   into a warm front

3. Which of these statements about *both* extratropical cyclones *and*
   hurricanes is *not* true?
   a. In the Northern Hemisphere,   c. They are low-pressure sys-
      air rotates counterclockwise   tems.
      around their centers.   d. They get a lot of energy
   b. Warm air is rising and pro-   from air mass interactions.
      viding a lot of precipitation.

4. A _____ is the line that marks the surface boundary
   between two air masses.

5. When colder air is advancing into space occupied by another air
   mass, its progress is marked on a weather map with what kinds of
   symbols?

6. Hurricanes develop along stationary fronts. (True or False)

7. When air masses begin to move, they tend to splinter into several
   smaller units. (True or False)

8. Why are extratropical cyclones associated with a lot of precipitation?

9. What has happened to the warm air in an extratropical cyclone system after occlusion has occurred?

10. Why don't extratropical cyclones form near the equator?

1. c	2. a	3. d	4. front
5. triangles or blue triangles	6. False	7. False	

8. Air is being lifted at the cold front, the warm front, and at the low-pressure center.

9. After the advancing cold front catches up to the warm front, all of the warm air will be lifted off of the ground, and the cold air mass will reunite with itself.

10. Extratropical cyclones form due to the interaction of differing air masses. Air masses at the equator are insufficiently different to produce those kind of storms.

Links to Other Chapters

- Varying air mass temperatures are a result of heating differences of places with different latitudes (chapters 1 and 2).
- Heavy rain or snow can fall due to the rising air in extratropical cyclones at fronts and at the low-pressure center (chapters 3 and 4).
- Low-pressure circulation is a precursor of the development of extratropical cyclones (chapter 4).
- The latitudes affected by extratropical cyclones vary over a year as the location of the subpolar low-pressure zone seasonally fluctuates (chapter 5).
- Air mass interactions and seasonal storms are important elements of midlatitude climates (chapter 7).
- Precipitation from extratropical cyclones and from tropical cyclones that have moved inland can cause runoff (chapter 12) and flooding (chapter 13).
- High winds from extratropical cyclones and tropical cyclones can lead to windblown sand on beaches (chapter 14) and the generation of large waves (chapter 15).

7 Climate

Objectives

In this chapter you will learn that:

- Climate is the weather that would be expected based on mostly global factors and some local factors.

- Latitude causes seasonal changes to insolation strength and length of day.

- Latitude affects which elements of the general circulation will be nearby.

- Movement of the subtropical high-pressure system will produce seasonal climate effects—particularly on west coast shorelines.

- Continentality will influence both the daily and yearly temperature range.

- Altitude can create a cold climate that would not otherwise be expected.

- Local terrain effects can influence climate.

Climate Elements

Climate is the weather you expect; it is average weather. It is not day-to-day variation (although that factor would be part of a place's climate). It is big-picture weather. A place's climate tells (1) how hot or cold it is, (2) how wet or dry it is, and (3) how those factors change over the course of a year.

Climate is produced by the interactions of the elements in all of the previous chapters: latitude, angle of insolation, seasons, daily rotation, the constituents of the atmosphere, continentality, pressure and wind, general circulation of the atmosphere, precipitation, and air mass interactions. If you consider climate hot-cold, wet-dry, and seasonal change, then a place's climate can be revealed by asking three fundamental questions about (1) latitude—seasonality, (2) latitude—general circulation, (3) continentality, and a contingent question about (4) terrain.

What are the three characteristics used to describe a place's climate?

_____

Answer: temperature, precipitation, seasons

Climate Question 1: Latitude—Seasonality

A place's latitude will affect how intense the insolation will be at different times of the year and how many hours of insolation there will be each day. The amount and strength of insolation will affect heating and therefore temperature.

- Equatorial locations have 12 hours of hot sunlight every day and very little seasonal change.

- Tropical locations have a summer (high-Sun) and winter (low-Sun) cycle, but it is hot all year with no major changes. (Think of the Caribbean, or Hawaii.)

- Midlatitude locations (e.g., the lower forty-eight United States) have more distinct summer and winter seasons.

- Arctic/polar locations have very distinct seasons. Cool summers with very long days turn into very cold winters with very long nights.

- Seasonal change in the general circulation: changing patterns of insolation over the course of a year will cause the elements of the general circulation to "follow the Sun" north or south. The elements of the general circulation move toward their hemisphere's pole in the summer and toward the equator in the winter.

How does latitude give a clue to the heating and temperature a place will have? _____

> *Answer:*   Latitude has a year-round affect on the amount of daylight and the strength of insolation that a place receives.

Climate Question 2: Latitude—General Circulation

The *up, down, up, down* and *east, west, east patterns* of air movement have a strong influence on temperature and precipitation, creating recognizable patterns:

- The intertropical convergence zone (ITCZ). This converging-wind area is cloudy and showery all the time.

- Trade winds. The reliable trade winds always blow from the east to the west. A place on an east-facing tropical coast can get a large amount of rain because the winds blow warm, humid air from the oceans onto the land.

- Subtropical high pressure (STH). This is a high-pressure zone, and descending warming air is going to generate blue skies and abundant sunshine (little rain) almost all the time.

- Westerlies. These reliable winds always blow from west to east. A place on a west-facing coast can get a fair amount of rain because the winds blow damp air from the ocean onto the land.

- Subpolar low. This is a low-pressure zone. Converging westerlies and polar easterlies bring contrasting air masses together. Extratropical cyclones and rainy, stormy weather are common.

How does the general circulation of the atmosphere give a clue to the kind of climate a place will have? _____

Answer: Different elements of the general circulation are characterized by fair skies or rainy weather. If prevailing winds come from over an ocean, then a place also can receive a lot of precipitation.

Seasonal Movement of Subtropical High Pressure

The seasonal movement of the ITCZ is relatively easy to understand. The ITCZ will be found near the place that is getting the most heat from the Sun (see the photo on page 84 of chapter 5). That intense heating will cause air to rise, which leads to a constant rainy/drizzly/thunderstorm kind of weather. If the ITCZ moves far away in the local winter, then its rainy qualities will move away, too.

The seasonal movement of the STH also has climatic implications, because the STH is the origin for the surface westerlies. The changes in position of the subtropical high-pressure region will be given special attention in this section because of the way it affects climate in the United States, particularly on the West Coast.

In the general circulation model, the STH near 30° latitude pushes surface air out, toward the pole and the equator. The air being pushed toward the pole will be turned by the Coriolis force to flow from west to east in the middle latitudes. The westerlies converge near latitude 60° and ascend in the stormy polar low-pressure zone there. The West Coast of the United States is in the westerlies zone between the STH and the subpolar low. Winds come off of the Pacific Ocean onto the shoreline. On the East Coast of the United States, prevailing winds blow across the continent before reaching the shore.

In the Northern Hemisphere winter, the ITCZ follows the Sun into the Southern Hemisphere. And the Northern Hemisphere's STH, westerlies, and subpolar low also will move southward. As the westerlies and subpolar low move south, they begin to have a greater influence on the weather of west coasts in the middle latitudes. As the STH moves away, it takes its high-pressure, fair weather with it. The westerlies will now blow more directly off the Pacific onto the U.S. West Coast. Wet maritime air and the subpolar low-pressure area are now the dominant weather influences, and the U.S. western shore has damp, rainy, stormy weather. Whereas the subpolar low is also affecting the U.S. East Coast, the westerlies still blow over land before reaching the ocean there, so the seasonal change is not as great there.

In the Northern Hemisphere summer, the Sun is at its farthest north of the equator. As the Sun heads north, it pushes the STH into a position to dominate weather in the midlatitudes. The westerlies are pushed farther toward the pole. The subpolar low is far away and has minimal summer influence on weather. Warm, dry, descending air from the STH is now brought into the West Coast of the United States. A sunny, dry period dominates the weather. On the U.S. East Coast, the

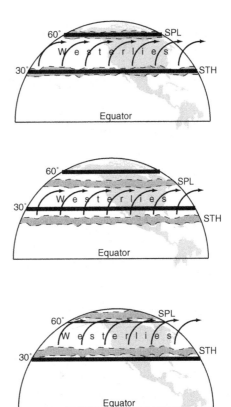

Figure 7.1. *Top:* Around the equinoxes, the position of the general circulation elements conforms to the model: The subtropical high (STH) is around 30° and the subpolar low is near 60°. On a continent's western coast, the westerlies blow off the sea and onto the north–south trending land in the high midlatitudes (the trend of the coast is N–S, not E–W, so the west wind blows straight on to it). On the eastern coast, the westerlies blow over land before reaching the shoreline. *Center:* In winter, the STH moves south, away from the midlatitudes, and its fair weather is replaced with the subpolar low. The eastern coast still has a drier, land-crossing wind. *Bottom:* In summer, the Sun has induced the northward movement of the STH, the westerlies, and the subpolar low. The STH comes to dominate weather in the midlatitudes as the subpolar low is pushed away.

displacement of the subpolar low changes the weather, but again, not as dramatically as on the other side of the continent.

What would be the rainy season on the West Coast of the United States? _____

Answer: Winter, when the STH is away to the south and the subpolar low exerts greater influence on weather.

Climate Question 3: Continentality

Continentality is an indication of how coastal or inland a place is. Places on an ocean shoreline have zero continentality (they are coastal). Places in the interior of a landmass, far from an ocean, would be considered to have a very continental location. Inland spots heat up more easily and cool off quicker than places near large water bodies. Inland places have hotter summers and colder winters. Inland places also have a greater daily temperature range than coastal places. Coastal places can get a lot of rain if they face into prevailing winds (trades or westerlies), which are loaded with moisture evaporated from the ocean.

On a hot summer day, why would it be cooler at an ocean beach than it is inland? _____

Answer: The ocean does not heat up as quickly as the solid landmass does, so in the summer the ocean is likely to be cooler than the land is.

An Exercise in Climate Determination

Answering these three questions—about: (1) latitude—seasonality, (2) latitude—general circulation, and (3) continentality—will explain almost everything about a place's climate. (More about the fourth question regarding terrain later in this chapter.) These three questions will help with the assessment of the basics of climate: temperature, precipitation, and seasonal change.

Answering the three questions about latitude and continentality will help to determine what kind of climate the eight imaginary cities on the imaginary Earth in Figure 7.2 will have.

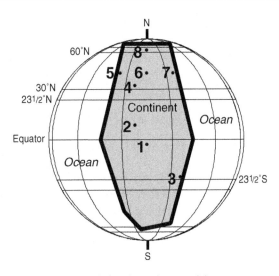

Figure 7.2. Eight imaginary cities on an imaginary Earth.

City 1

Question 1: Equatorial—it's about the same all year, 12 hours of hot sun every day

Question 2: ITCZ—low pressure

Question 3: Continental (but continentality is not so important at the equator because every day is about the same)

Climate: Hot all year, rain and thunderstorms all year, especially when the ITCZ is near

City 2

Question 1: Tropical—not very strong seasonal differences

Question 2: Trade winds and seasonal switches of the ITCZ

Question 3: Continental

Climate: Hot all year, dry and rainy seasons as the ITCZ moves north and south

City 3

Question 1: Southern Hemisphere, tropical—slight seasonal differences

Question 2: Trade winds blow from east to west

Question 3: Coastal (facing into the trade winds)

Climate: Warm all year with a hot summer, wet all year as winds blow humid ocean air inland, the STH will approach and dry things up a bit during the Southern Hemisphere winter

City 4

Question 1: Midlatitude 30°—starting to see seasons develop

Question 2: Subtropical high pressure

Question 3: Continental

Climate: Hot summer when sunlight is most direct, dry all year because of the STH, deserts

City 5

Question 1: Midlatitude—clear seasons

Question 2: Westerlies blow from west to east

Question 3: Coastal (facing into the westerlies)

Climate: Ocean moderates temperature changes, cool summer, warm winter, winter westerlies blow damp air inland—lots of rain, nearby STH brings summer dryness

City 6

Question 1: Midlatitude—clear seasons

Question 2: Westerlies

Question 3: Continental

Climate: Hot summer, cold winter, rain or snow from air mass interactions

City 7

Question 1: Midlatitude—clear seasons

Question 2: Westerlies blow from west to east

Question 3: Coastal (westerlies come from the land side)

Climate: Ocean moderates temperature changes, warm summer, cool winter, rain or snow from air mass interactions; STH not as important on east coasts as on west coasts

City 8

Question 1: High latitude—cold with very strong summer/winter differences

Question 2: Subpolar low switching with polar high pressure

Question 3: Continental

Climate: Cool summers, very cold winters, might be low-pressure storms—but it may be that the air is too cold to hold much moisture

Why do cities 5, 6, and 7 have such different climates despite being at the same latitude? _____

Answer: City 6 is continental, so it will have greater temperature extremes than coastal cities 5 and 7. City 5 is a West Coast city in the westerlies. City 7 is on an east-facing coast.

Climographs

A place's climate is often depicted with a tool called a climograph. A climograph shows how temperature and precipitation vary over the course of twelve months.

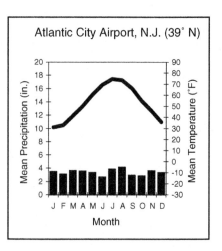

Figure 7.3. This is a climograph for the airport at Atlantic City, New Jersey. In a typical January (which is the year's coldest month), the average monthly temperature is 30.9°F and there will be 3.46 inches of precipitation. In a typical July (the hottest month), the average monthly temperature is 74.7°F and there will be 3.83 inches of precipitation. In this midlatitude, East Coast location there is no seasonal pattern to precipitation quantity.

The twelve months are the normal time intervals shown on a climograph. The months run along the x-axis, usually in calendar-year order: January, J, to December, D. (Sometimes Southern Hemisphere cities are shown from July to June so that their December, January, February summer is plotted in the middle of the graph.) Average monthly precipitation is marked on one y-axis. Bar graphs are used to tell how much precipitation falls in an average month. Snow is melted to derive a liquid measurement. Average monthly temperature is marked on the second y-axis. A continuous line shows the average temperature from month to month.

Use Figure 7.3 to determine the mean annual temperature range at Atlantic City, New Jersey. _____

Answer: Precision isn't critical, but it is 43.8°F from January to July.

Real Climates

Back in the real world, there are eight locations that correlate to the cities from the imaginary Earth whose climates were described earlier in this chapter. The answers to the three climate questions for the imaginary cities also apply to their real-world counterparts, as shown by the climographs in Figures 7.5 and 7.6.

The following are interpretations of the climate for the eight cities shown in Figure 7.4.

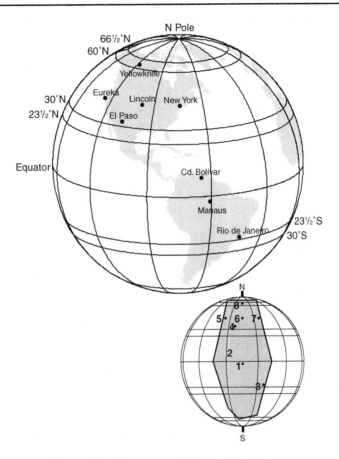

Figure 7.4. There are eight real cities whose climate situation corresponds with the eight imaginary cities. Inset from Figure 7.2.

Yellowknife, Canada (City 8)

High latitude, very seasonal. Cool summers, very cold winters. Little precipitation because the air is too cold to hold much moisture.

Eureka, California (City 5) ; Lincoln, Nebraska (City 6) ; New York City (City 7)

Seasonal contrasts. Continental Lincoln has greater temperature extremes than the two coastal cities. Eureka's dry summer is from the north-moving STH; its wet winter is from the marine westerlies and the subpolar low.

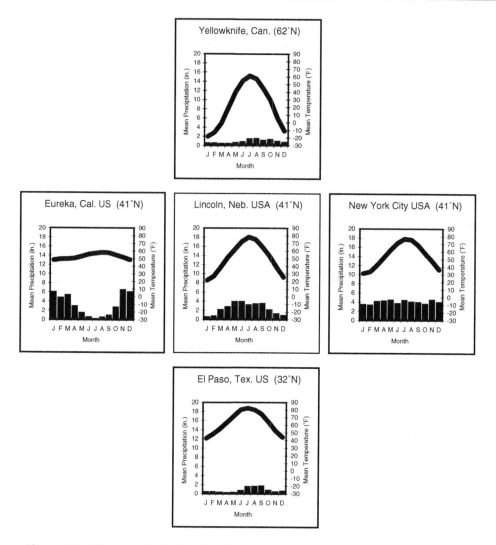

Figure 7.5. Climographs for the five North American cities shown in Figure 7.4.

El Paso, Texas (City 4)

Continental location around 30° latitude. Seasonality is apparent in temperatures. Dry because of persistent high-pressure sunshine.

Ciudad Bolívar, Venezuela (City 2)

Hot all year long. Two temperature peaks, in April and September, when the sunlight is strongest. Precipitation seasons are caused by the summer approach and winter distance of the ITCZ.

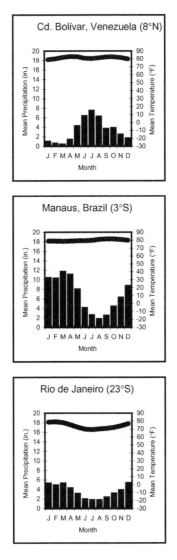

Figure 7.6. Climographs for the three South American cities shown in Figure 7.4.

Manaus, Brazil (City 1)

Equatorial location gives same hot temperature all year. Rain all year: least when the ITCZ is farthest north, from June to October.

Rio de Janeiro, Brazil (City 3)

The Sun is directly overhead in December at 23°S. January and February are hot. There is moderate seasonality. Trade winds bring rain all year long—least in the Southern Hemisphere winter, when the STH is farthest north.

Climate Question 4: Terrain

The fourth question is terrain. Sometimes a place is at such high altitude that the effects of the elevation (primarily on temperature) overwhelm the latitude and continentality influences and produce a distinct highlands climate.

Sometimes a high mountain range can interfere with the wind patterns of the general circulation of the atmosphere. Also, wet slopes or rain shadows can be produced depending on which way the wind is blowing and how a mountain range intercepts the prevailing wind.

Rough terrain also can affect the amount of sunshine that falls at a given place. South-facing slopes receive much more sunshine than north-facing ones in the midlatitudes. Areas in mountain valleys also might be blocked from sunlight.

How would temperatures in a mountainous area be different from temperatures at a nearby lower elevation? _____

Answer: It will almost always be colder at higher elevations.

SELF-TEST

1. A climograph will show the temperature for the hottest day of the year. (True or False)

2. A climograph will show how much rain falls on the wettest day of the year. (True or False)

3. A climograph will show the average number of rainy days each month. (True or False)

4. A climograph will show the average temperature for each month. (True or False)

5. A climograph can be used to tell how much the temperature is likely to change from month to month. (True or False)

6. The wet/dry season in Eureka, California (lat. 41°N on the Pacific coast) is caused by movement of the _____.

7. The wet/dry season in Manaus, Brazil (close to the equator at 3°S) is caused by movement of the _____.

8. The temperature difference between a typical February day in Rio de Janeiro and New York City is approximately _____. (Use the climographs in this chapter.)

9. Why is it relatively cold in Rio de Janeiro in June, July, and August?

10. What are the four questions to ask to determine what a place's climate will be like?

ANSWERS

1. False   2. False   3. False   4. True   5. True

6. STH *or* subtropical high-pressure zone

7. ITCZ *or* intertropical convergence zone

8. 40°–50°F (79.2°F in Rio and 33.6°F in New York = 45.6°F difference)

9. Because Rio is in the Southern Hemisphere and it is winter in those months.

10. 1. What is its latitude (seasons)?

    2. What is its latitude (general circulation of the atmosphere)?

    3. Is it coastal or continental?

    4. Does its terrain or elevation exert a special influence?

Links to Other Chapters

- Latitude, angle of insolation, seasons, daily rotation, the constituents of the atmosphere, continentality, pressure and wind, general circulation of the atmosphere, precipitation, and air mass interactions are the building blocks of climate, and they are addressed in chapters 1 through 6.
- Physical and chemical weathering processes are greatly dependent on the availability of water or ice (chapter 11).
- The amount of water in the soil is a function of how much evaporation and precipitation there are at a place (chapter 12).
- Stream flow and flooding are determined by the quantity and timing of precipitation and melting within the stream's watershed (chapter 13).
- Deserts have a climate that is too dry to support plants. The lack of vegetation allows the action of the wind to be an important geomorphic process (chapter 14).
- Places that are so cold that more snow falls in the winter than can melt away in the summer will be where glaciers can be found (chapter 14).

8 Plate Tectonics

Objectives

In this chapter you will learn that:

- Geologic time means that there is enough time for anything that could happen to happen.
- Volcanoes, earthquakes, mountains, island chains, midocean ridges, and deep-ocean trenches exist in worldwide patterns—they are not randomly distributed.
- The outer layer of the solid Earth is the crust.
- There are two types of crust: a thick continental crust and a thinner, denser oceanic crust.
- Earth's outer portion is made up of independent sections called plates, which are composed of the upper, rigid part of the mantle, with a covering of crust.

Geologic Time

Geologic time is an idea that refers to the age of Earth—4 billion to 5 billion years. For our purposes, geologic time means: There is enough time for anything that could happen to happen.

If some process needs 1,000 years to run its course—okay, Earth can wait. If something takes 500,000 years—no problem, there have been ten thousand 500,000-year runs since Earth came into being. The concept of geologic time means that a process can have as much time as it needs to be effective. A stream cutting into rock at the rate of 1 mm a year seems awfully slow. But, 1 mm a year for 1 million years would cause 1 km of erosion. If Africa and Europe are splitting apart from North and South America at the rate of 3 cm each year, they would be 6,000 km apart after 200 million years.

Much of what is discussed in geomorphology depends on the idea of geologic time. Ideas and concepts will be presented in chapters 8 through 15 (but especially in chapters 8 through 11) as though they happen instantaneously or from year to year. Often there are many, many cycles or events that produce the effects being described. A stratovolcano with alternating ash and rock layers can take lots of eruptions to create the high, snowcapped, cone-shaped forms we recognize. When the freezing of water is said to split rocks, that may be true, but it might take thousands of freeze-thaw cycles to bring the rock to its breaking point. Nevertheless, geologic time is available. If it takes a thousand years or a thousand cycles, then that's what it will take. Earth doesn't care.

This canyon and Rainbow Bridge, Arizona, were produced by the very slow but persistent erosive effects of moving water. (Image courtesy of National Geophysical Data Center.)

The concept that Earth processes can happen regardless of how much time it would take to produce them is _____.

Answer: geologic time

Random and Systematic Distributions

The arrangement of items in a space can be described as a random distribution or a systematic distribution. If there is a random arrangement of items, the location of any one object has no relationship to the location of any other object. The item is where it is by pure chance alone. If the distribution of objects is not random, it is said to be systematic. In a systematic distribution, objects are more likely to be in particular locations; it is not just chance that determines position. If there is a

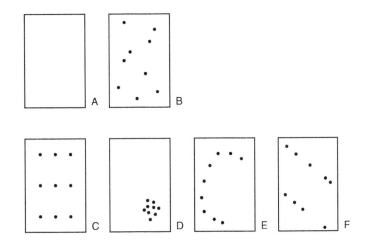

Figure 8.1.   Consider a blank map of a place (Map A). Map B shows the location of nine objects (they could be anything: volcanoes, earthquake epicenters, fast-food restaurants, houses where college students live, etc.). If the distribution of those nine objects is completely random—that is, if an object has an equal chance of being anywhere on the map—then the distribution might look as in Map B. Maps C to F also show a distribution of nine objects in space. However, these are not random arrangements. Objects are not as likely to be anywhere on these maps as anywhere else. In Map C, there is a very regular grid layout. Objects are in rows and columns, and only in rows and columns. In Map D, the nine objects are all in a cluster in one section of the map. In Map E, the nine objects are throughout the space, but they appear to be in a semicircular arrangement. Likewise in Map F, the nine objects are arranged in two parallel lines.

nonrandom arrangement, then very likely something is causing items to be located in particular places.

What could it mean if a phenomenon has a systematic spatial arrangement? _____

Answer: If something is systematically arranged, it is likely that there is some force acting to cause that distribution.

Distributions of Geographic Phenomena

There are many phenomena in physical geography that are found throughout the world for one reason or another. Mapping their occurrences will show that they are not likely to be anywhere; they are found in particular places. Glaciers are not randomly distributed, as you are not as likely to find a glacier in Florida as in Alaska: glaciers need sustained cold temperatures. Similarly, a particular plant such as a palm tree is not going to be randomly found around Earth. The location of palm trees is restricted by climatic forces: there is a reason they are found where they are and why they are not found in other places.

In the world of geomorphology and Earth mechanics, there are a number of landscape elements or activities whose worldwide distribution suggests that they are not randomly located around the planet. Geomorphic forms such as mountains, islands, midocean ridges, and deep-ocean trenches are located in very particular kinds of places. These landforms are not found just anywhere on the continents or anywhere on the seafloor. Likewise, strong earthquakes and volcanoes also are found in particular places. Their locations seem to be systematically determined, too.

The worldwide pattern of mountains shows that these landforms are likely to be found in strings, or chains, that are usually toward one side of a continent (not in the middle).

The worldwide pattern of oceanic islands shows that these features are likely to be found in chains or long, linear patterns.

The worldwide pattern of midocean ridges shows that these underwater mountains are found in long, linear patterns in the middle of the sea, not near the continents. They also seem to run parallel to the shorelines of the continents on either side of the ocean basin.

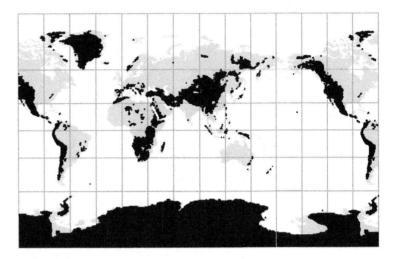

Figure 8.2. Places that have elevations of at least 1,000 meters (3,300 feet) are indicated in black. (The areas of high elevations in Antarctica and Greenland are largely due to the thickness of glacial ice; moreover, their extent is greatly exaggerated because of the way this cylindrical projection distorts high latitudes.)

The worldwide pattern of deep-ocean trenches shows that these deep, linear, underwater valleys are only found adjacent to continents or island chains. The trenches always run parallel to the neighboring land.

The worldwide pattern of strong earthquakes shows that earthquakes tend to be found in linear chains or in isolated clusters. These

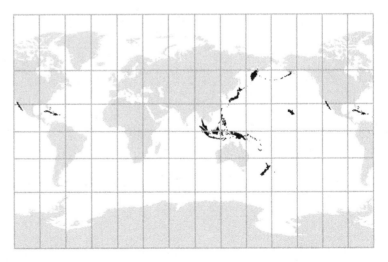

Figure 8.3. Oceanic islands tend to occur in long, linear chains.

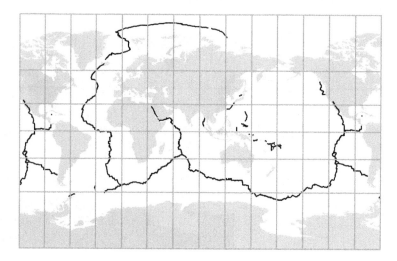

Figure 8.4. The linear pattern of midocean ridges.

chains tend to be found in three special kinds of places: (1) along the edge of continents, (2) along island chains, or (3) right down the middle of an ocean.

The worldwide pattern of volcanoes shows that volcanoes tend to be found in linear chains or in isolated clusters. The chains tend to run parallel to the edge of continents and a bit inland from the water. Volcanoes also run along island chains. They also can be found in isolated clusters called hot spots.

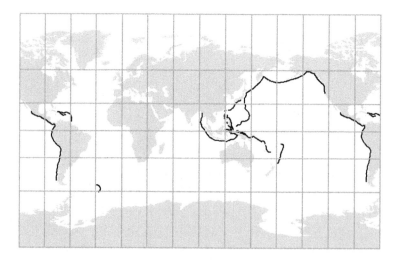

Figure 8.5. The locations of deep-ocean trenches.

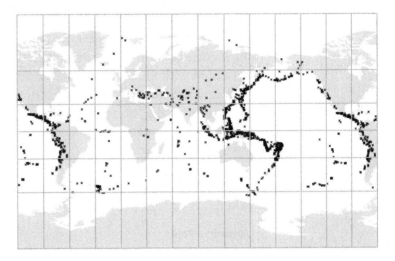

Figure 8.6. The epicenters of 1,561 earthquakes with magnitudes of at least 6.0 in the ten years from January 1991 through December 2000. (Data from USGS National Earthquake Information Center.)

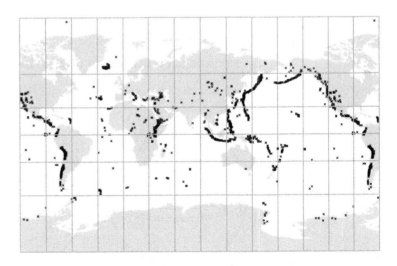

Figure 8.7. The worldwide distribution of 1,521 volcanoes that have been active in the past ten thousand years. (Data from Smithsonian Institution, Global Volcanism Program.)

The patterns of landforms and volcanic activity and earthquakes suggest systematic cause. The plots in Figures 8.2 through 8.7 show that there is a high correlation in the location of all these features. They all appear to be associated with the middle of oceans and with the edges of continents.

What could cause all of these features to appear in long, linear patterns along the edge of continents or with island chains? The need to explain these systematic occurrences will require an understanding of the structure of Earth.

Earthquakes and volcanoes are randomly distributed around Earth. (True or False) _____

Answer: False

Layers of Earth

Earth is a spherical-shaped planet that is made up of many layers: a two-layer core; an intermediate mantle; and an outer, solid layer called the crust. The crust runs about 5 to 25 miles in thickness. It is thickest under large continental mountains and thinnest at the bottom of the sea. Everything that people do that affects the solid Earth happens to the crust. There are two kinds of crust:

1. oceanic crust, which is relatively thin and made of very dense rock

2. continental crust, which is thick and made of less dense rock

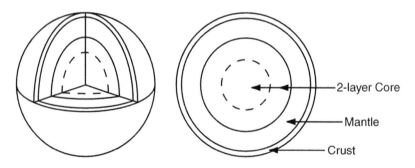

Figure 8.8. Earth is a spherically shaped planet made up of a two-layer core, the mantle, and the crust.

Oceanic crust forms when molten rock (magma) cools quickly underwater. Continental crust forms when magma cools slowly atop dry land, or within the crust itself.

The mantle is the layer below the crust. It is approximately 1,800 miles thick and is made up of three sections. The top part of the mantle is rigid, solid rock. The bottommost section of the mantle also is rigid, solid rock. The section between is soft and can be distorted or pushed around. It is called the asthenosphere.

How is oceanic crust different from continental crust? _____

Answer: Oceanic crust is not as thick and is denser (heavier).

Asthenosphere and Lithosphere

The crust and the upper part of the mantle are taken together and called the lithosphere. The lithosphere is the solid, rigid, outer portion of Earth. However, the lithosphere is not all one complete shell; it is made up of many pieces called plates. The lower part of each plate is made up of a section of the rigid upper mantle, and plates can be covered with either oceanic crust or continental crust (or with both). The two-layer lithosphere plates rest on top of the softer, distortable section of the mantle called the asthenosphere.

The asthenosphere is the key to plate tectonics. Because the asthenosphere is distortable, the overlying lithosphere plates can move: the lithosphere is not cemented to the solid rock of the lower mantle.

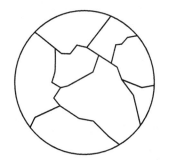

Figure 8.9. The entire surface of Earth is covered with lithospheric plates that can move independently of each other.

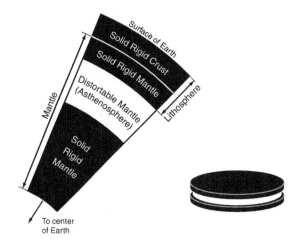

Figure 8.10.   *Left:* The mantle is the section of Earth that is beneath the rigid outer crust and is made up of three sections: a rigid upper mantle, a distortable middle (asthenosphere) section, and a solid, rigid lower mantle. *Right:* The lithosphere (crust plus upper, rigid mantle) and the lower, rigid section of the mantle sandwich the distortable asthenosphere.

Why can plates move? _____

> *Answer:*   Each plate section of the lithosphere is independent of other sections, and plates rest atop a soft, deformable mantle section called the asthenosphere.

SELF-TEST

1. The outermost layer of the solid Earth is the _____.
   a. asthenosphere
   b. outer core
   c. mantle
   d. crust

2. What is the layer of the lithosphere that is below its top crust?
   a. asthenosphere
   b. upper mantle
   c. outer core
   d. inner core

3. The independent sections of the lithosphere are called
   _____.

4. Earth is approximately _____ years old.

5. In a systematic distribution there is some force that is acting to influence where objects or phenomena appear. (True or False)

6. Mountains are randomly distributed across Earth. (True or False)

7. A plate can be covered with continental crust, oceanic crust, or with sections made of either type of crust. (True or False)

8. Why is geologic time such a powerful concept?

9. What two kinds of locations are good places to look for volcanoes?

10. Why is oceanic crust different from continental crust?

ANSWERS

1. d   2. b   3. plates   4. 4 billion to 5 billion years

5. True   6. False   7. True

8. It means that even if a process takes a very, very long time, it still can be effective. There is enough time for anything that could happen to happen.

9. The locations are near the edge of a continent and in the middle of an ocean.

10. Oceanic crust is formed when magma cools quickly underwater. Continental crust is formed when magma cools slowly within the Earth or when exposed to the air.

Links to Other Chapters

- The landforms created by plate interactions depend on the type of crust that covers the plate (chapter 9).
- When oceanic crust plates and continental crust plates collide, there are shorelines, trenches, and volcanoes (chapter 9).
- When oceanic crust plates collide, there are volcanic island chains and trenches (chapter 9).
- When continental crust plates collide, there are mountain chains (chapter 9).
- When oceanic crust plates diverge, there are midocean ridges and volcanoes (chapter 9).
- Earthquakes are associated with plate boundaries because of stress and strain put on the crust in those places (chapter 10).
- Whether a volcano is explosive depends on the chemical properties of its magma (chapter 10).
- Layers of Earth and different types of crust will help provide evidence for the theory of plate tectonics (chapter 10).
- Evidence for the theory of plate tectonics is listed near the end of chapter 10.

9 Plate Interactions

Objectives

In this chapter you will learn that:

- The outer, solid layer of Earth is made of sections called plates that are free to move independently of one another.

- When plates move, they get either closer or farther from neighboring plates.

- Convergence involving a plate with oceanic crust causes subduction, which leads to the creation of deep ocean trenches and explosive volcanoes.

- A collision of two continental crust covered plates will unite the two plates and create a mountain chain along their old border.

- When plates split apart, new ocean crust is added to each plate, and a midocean ridge will form at the spreading boundary.

- Earth's physical features can be used to diagnose what type of plate interactions are occurring in particular places.

Lithospheric Plates

The lithosphere is the outer, rigid section of Earth. The lithosphere is made up of many pieces, called plates. The lower part of each plate is composed of a section of the rigid upper mantle. Plates can be covered with either oceanic crust or continental crust (or with separate sections of both types). There are many different plates on Earth, and each can move independently of the others, although its movement will affect the neighboring plates. Each of the tectonic plates can move independently of the other plates because it sits on top of the flexible asthenosphere.

Remember that lithospheric plates are two-layer systems. The bottom layer is the rigid, upper part of the mantle. When we refer to an object as a "continental crust plate," that really means a bottom mantle layer with continental crust atop it.

If a plate moves, three things can happen:

1. There will be a collision with an adjacent plate.

2. A gap will open between it and an adjacent plate

3. The plate will slide past its neighbor without getting closer or farther apart.

Three types of collisions can occur between plates:

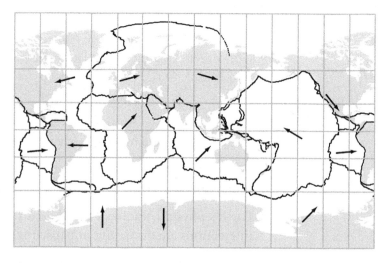

Figure 9.1. Plate boundaries and general directions of movement for the major plates.

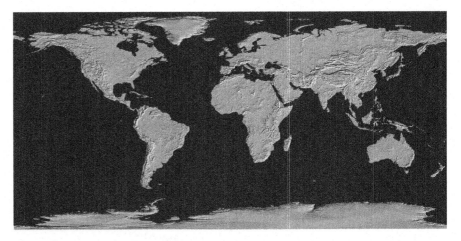

This shaded relief map of the world shows the surface features of the continents. You can see how shorelines, hills and mountains, and island chains are correlated with boundaries of tectonic plates. (Globe Project image courtesy of National Geophysical Data Center.)

1. A continental crust plate collides with an oceanic crust plate: C→ ←O.

2. An oceanic crust plate collides with an oceanic crust plate: O→ ←O.

3. A continental crust plate collides with a continental crust plate: C→ ←C.

Three types of plate separations can occur:

1. An oceanic crust plate separates from an oceanic crust plate: ←O O→.

2. A continental crust plate separates from a continental crust plate: ←C C→.

3. A continental crust plate separates from an oceanic crust plate: ←C O→.

Plates move very slowly—a few centimeters per year—so these collisions and separations do not occur rapidly. In the rest of this chapter we focus on the six types of plate interactions that result from convergence (collisions) or divergence (separations). What is depicted in a few sketches may take millions of years.

What are the three ways in which a plate can move with respect to an adjacent plate? _____

Answer: Plates can converge, diverge, or slide past each other.

Continental–Oceanic Crust Plate Collisions

Case 1: A plate covered with continental crust collides with a plate covered with oceanic crust: C→ ←O.

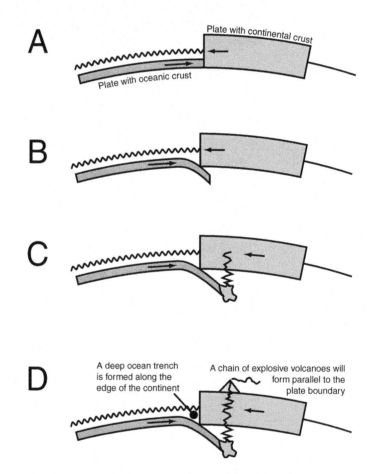

Figure 9.2. *A:* When an oceanic crust plate collides with a continental crust plate, the bigger, thicker continental crust plate will win the collision. *B:* The continental crust plate will force the oceanic crust plate to subduct. *C:* As the oceanic crust is forced deep into Earth, the crust will melt, and that magma (molten rock) will begin to work its way back up, through the continental crust covered plate, to the surface. *D:* If the magma remains molten all the way up, it can escape to the surface through volcanic eruptions.

The bigger, thicker continental crust wins this collision. The continental crust plate will run over and sink the heavy oceanic crust plate. As the collision progresses, the oceanic crust plate will be forced downward, below the continental crust plate, in a process called subduction. As a result of the collision and subduction, a deep ocean trench forms parallel to the shoreline of the continent. When the oceanic crust is forced deep into Earth, the crust melts, and that magma (molten rock) will begin to work its way back up, through the continental crust-covered plate, to the surface. The magma will acquire some of the properties of the surrounding continental crust as it oozes through it.

If the magma works its way up, it can escape to the surface through volcanic eruptions. The volcanoes would be located parallel to the shoreline, a bit inland from the edge of the continent. Continental–oceanic crust plate collisions are indicated by deep ocean trenches parallel to the edge of a continent and by a chain of mountains and volcanoes a bit inland from the shoreline.

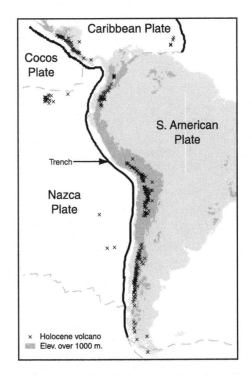

Figure 9.3. The West Coast of South America is a place where a continental crust covered South American plate is colliding with an oceanic crust covered Nazca plate. Note the shore-parallel trench just off the coast and the string of volcanoes parallel to the shoreline and a bit inland.

Why does the oceanic crust plate subduct in this type of collision?

_____

Answer: The thicker continental crust plate will cause the thinner, heavier, oceanic crust plate downward.

Oceanic–Oceanic Crust Plate Collisions

Case 2: A plate covered with oceanic crust collides with a plate covered with oceanic crust: O→ ←O.

   In this evenly matched collision, one plate or the other will prevail, and the losing plate will be subducted. As the subducting plate gets deeper into Earth, the plate will melt, and that magma will work its way upward to the surface. If the molten material from the subducting plate works its way through to the top of the other oceanic crust plate, that magma can escape out to the seafloor through underwater volcanoes. If

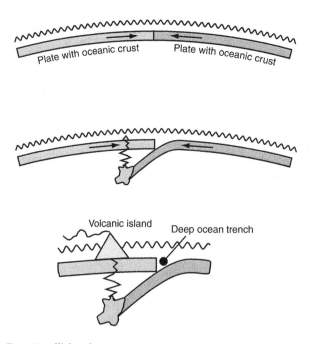

Figure 9.4. *Top:* A collision between two oceanic crust plates. *Center:* One or the other will prevail, and the losing plate will be subducted. As the subducting plate gets deeper into Earth, the plate will melt, and that magma will work its way up to the surface, where it can emerge in volcanic eruptions. *Bottom:* If an underwater volcano builds enough material for the feature to pierce the surface of the sea, it will create an island—or a string of islands parallel to the collision boundary.

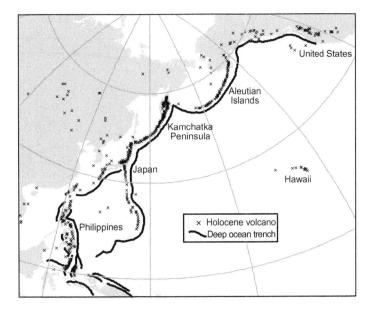

Figure 9.5. The Philippines, Japan, the Kamchatka Peninsula, and the Aleutian Islands are places where plates covered with oceanic crust are colliding. Note the strings of oceanic volcanic islands that run parallel to deep ocean trenches.

an underwater volcano builds enough material to pierce the surface of the sea, it will create an island—or a string of islands parallel to the collision boundary. A deep ocean trench also will develop, parallel to the subduction boundary, and parallel to the island chain. Oceanic–oceanic crust plate collisions are indicated by chains of volcanic islands and an adjacent, parallel deep ocean trench.

How would you know to suspect that a collision between two oceanic crust plates has taken place? _____

Answer: A chain of volcanic islands that runs parallel to an adjacent deep ocean trench would be evidence of that type of collision.

Continental–Continental Crust Plate Collisions

Case 3: A plate covered with continental crust collides with a plate covered with continental crust: C→ ←C.

In this type of plate collision, there is no subduction because each type of crust is of relatively low density and therefore cannot sink down

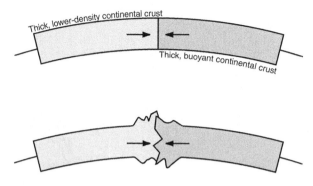

Figure 9.6. *Top:* When two plates with continental crust collide, there is no subduction. *Bottom:* The collision just progresses, and the two unsinkable plates mangle, warp, and crumple each other.

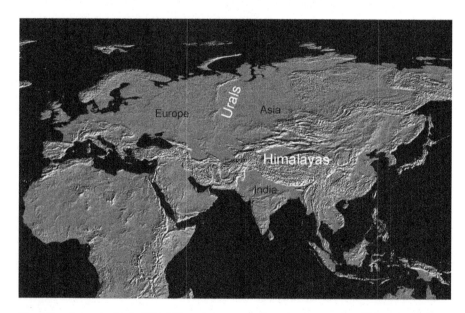

The Himalayas in India/Tibet/Nepal are an active continental collision boundary. India, on one plate, is moving north and colliding with the Eurasian plate. The Ural Mountains in Russia are a very old range that mark an ancient collision boundary. Prior to their collision, Europe and Asia were on different plates. Those plates were fused together in the continent–continent crash that produced the Urals. (Globe Project image courtesy of National Geophysical Data Center.)

into Earth. (It's like trying to sink a life preserver.) Because there is no subduction, the two unsinkable plates mangle and warp and crumple each other. This will join the two plates together and produce a mountain range along the collision boundary. Continental–continental crust plate collisions fuse the two colliding continents together and produce a mountain chain along the impact boundary.

Why is there no subduction when two continental crust plates collide?

Answer: Continental crust is thick and of low density, and that type of plate is unable to subduct into the asthenosphere.

Oceanic Crust Covered Plate Splits

Case 4: A plate covered with oceanic crust splits apart (or separates) from a plate covered with oceanic crust: ←O O→.

If a plate covered with oceanic crust is split in two pieces, the pieces will move apart from each other. As the two pieces separate, magma

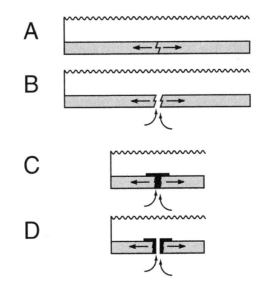

Figure 9.7.   _A:_ A plate covered with oceanic crust begins to break, and _(B)_ the pieces move apart. _C:_ Magma (molten rock) flows upward into the gap between the sections. As the magma comes into contact with the cold water, the magma will cool to form new oceanic crust. _D:_ If the splitting continues, another gap can open, and more magma can escape.

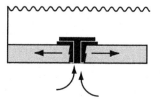

Figure 9.8. Over time, these split-fill cycles of plate separation and magma emergence can create an underwater mountain range.

flows upward into the gap that opens up. The magma flows out onto the ocean floor. As it comes into contact with the cold water it cools quickly and forms new oceanic crust. If the splitting continues, another gap can open, and more magma can escape to create an additional section of new oceanic crust at the breach. Many, many cycles of splitting–magma escape–cooling can produce an accumulation of new oceanic crust along the splitting boundary. After numerous cycles, an elevated area called a midocean ridge can be created; it looks like an underwater mountain chain along the rift zone.

Figure 9.9. Seafloor spreading due to divergence of two oceanic crust plates is occurring in the middle of the Atlantic Ocean. Note how the Middle-Atlantic Ridge runs down the middle of the ocean basin, and how its orientation matches the general shapes of North and South America on one side, and Europe and Africa on the other side.

The oceanic crust closest to the splitting will be the newest on the seafloor. The farther the crust is from the splitting zone, the older it will be. A midocean ridge is evidence that an oceanic crust plate has split into two diverging sections.

What is the geomorphic evidence that indicates an ocean is getting wider? _____

Answer: A midocean ridge, produced by seafloor spreading.

Continental Crust Covered Plate Separates

Case 5: A plate covered with continental crust splits apart (separates) from a plate covered with continental crust: ←C C→.

If some tectonic force breaks apart a landmass, the two new continental crust plates will begin to move apart from each other. When a

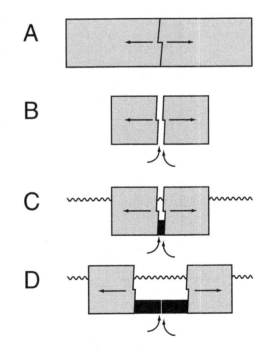

Figure 9.10. *A:* As a landmass breaks apart, *(B)* the two new continental crust plates will begin to move apart from each other. *C:* When a gap opens up in the land, water will move in to fill up the space as molten material from below tries to work its way upward into the gap. If magma escapes, it will cool rapidly underwater to form new oceanic crust. *D:* Eventually this sequence can be transformed into a case of seafloor spreading (case 4).

gap opens in the land, either (1) the ocean will move in to fill the space, or (2) it will create a valley that will fill with water and become a lake. At the same time, molten material from below will try to work its way upward. If the molten rock escapes, it will cool rapidly underwater, and new oceanic crust will be created in the rift zone. As the splitting continues, the situation transforms itself from continents diverging to a case of seafloor spreading (case 4).

Notice how the splitting of a continent into two pieces will create a situation where oceanic crust is added onto the edges of the two continental crust plates. If there are two matching puzzle-piece continent landmasses on either side of an ocean and there is a correspondingly shaped midocean ridge precisely between them (Figure 9.9), that is a good indication that the ocean was formed when a large landmass broke apart.

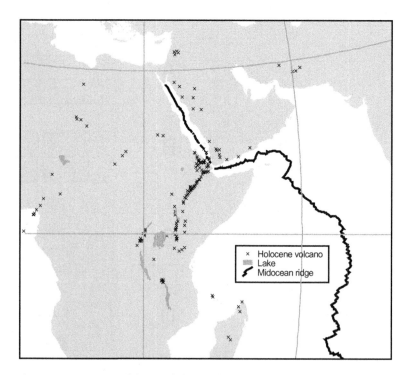

Figure 9.11. East Africa and the Red Sea are regions where a continent is currently splitting into two diverging sections. Note the long, linear lakes in East Africa and their association with strings of places that have had volcanic eruptions. In Northeast Africa, the split with the Arabian Peninsula has progressed to the point where midocean ridges have formed in the narrow sea.

Why is the new crust that forms between sections of a splitting continent the oceanic type of crust and not the continental type of crust?

Answer: The new crust is formed when magma cools rapidly underwater, so it is the oceanic type.

Oceanic–Continental Crust Plate Separates

Case 6: A plate covered with oceanic crust separates from a plate covered with continental crust: ←C O→.

If a plate covered with both continental crust and oceanic crust broke apart on that boundary, this would quickly (geologically speaking) turn into a situation like the preceding one, where new oceanic crust forms in the gap and becomes part of each of the two diverging plates. A case like this is largely theoretical.

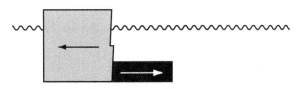

Figure 9.12. If a continental crust plate were to diverge from an adjacent plate with oceanic crust, it would transform into a situation where new oceanic crust forms in the gap and becomes part of each of the two diverging plates.

Diagnostic Geomorphology and Crustal Changes

Oceanic crust is destroyed in the subduction process that results from collision (cases 1 and 2), and new oceanic crust is created when plates spread apart from each other (cases 4, 5, and 6). Oceanic crust is created whenever magma cools quickly in water.

Continental crust is neither created nor destroyed by the process of plates moving around. Continental crust can be constructed when magma cools slowly within the solid part of Earth, or if it escapes at volcanoes and cools in air atop the ground.

Summary of Crust Changes and Landforms from Plate Interactions

Case	Description	Landform Evidence	Crust
1	Collision C→ ←O	Volcanoes on land a bit in from the coast Deep ocean trench parallel and adjacent to the landmass	Oceanic crust destroyed Additions to continental crust
2	Collision O→ ←O	A chain of underwater volcanoes or volcanic islands A deep ocean trench adjacent to the volcano string	Oceanic crust on one plate is destroyed Continental crust created if volcanoes breach the sea surface
3	Collision C→ ←C	Deeply distorted terrain A midcontinent mountain range	No change to quantities
4	Separation ←O O→	Midocean ridge Puzzle-piece continents on opposite sides of the ocean	Oceanic crust is created
5	Separation ←C C→	Long, linear seas, or deep, narrow lakes Midocean ridge if splitting continues	Oceanic crust is created
6	Separation ←C O→	Midocean ridge if splitting continues	Oceanic crust is created

Old continental crust is destroyed by weathering and erosion: How is new continental crust created? _____

Answer: New continental crust is created when magma cools slowly within Earth or if magma escapes at the surface and cools into new rock on top of Earth.

SELF-TEST

1. Which type of plate interaction neither creates nor destroys crust?
   a. Continental crust collides with continental crust:

      C→ ←C
   b. Oceanic crust collides with continental crust: O→ ←C
   c. Continental crust diverges from continental crust:

      ←C C→
   d. Oceanic crust diverges from oceanic crust: ←O O→

2. Which type of plate interaction can ultimately result in the creation of new continental crust?

a. Continental crust collides with continental crust:
C→ ←C

b. Oceanic crust collides with continental crust: O→ ←C

c. Continental crust diverges from continental crust:
←C C→

d. Oceanic crust diverges from oceanic crust: ←O O→

3. The process in which one plate is forced deep within Earth by its collision with another plate is called _____.

4. The underwater mountain ranges along the center of the oceans are called _____.

5. An example of a place where a continent is currently splitting apart is _____.

6. A continental crust plate *cannot* be forced downward into Earth when it collides with another plate. (True or False)

7. A deep ocean trench is likely to be near a parallel chain of volcanoes. (True or False)

8. Explain how a feature such as the Atlantic Ocean could be created by the rupture of a continental crust covered plate.

9. Why does a deep ocean trench indicate that a collision is taking place?

10. What evidence is produced to show that a plate covered with oceanic crust has collided with a plate covered with continental crust?

ANSWERS

1. a   2. b   3. subduction   4. midocean ridges

5. East Africa or the Red Sea   6. True   7. True

8. As a continent is split apart, new oceanic crust will form in the gap. As the spreading continues, a larger ocean basin will form, with a midocean ridge at its center.

9. It is a sign that an oceanic crust covered plate is subducting, which occurs when plates are converging.

10. A deep ocean trench will form adjacent and parallel to the continent. There can be a chain of volcanoes parallel to and a bit inland from the shoreline.

Links to Other Chapters

- Plate interactions happen on the geologic time scale (chapter 8).
- Plate movement is allowed by the structure of Earth's mantle and crust layers (chapter 8).
- Types of volcanic eruptions are correlated with their locations on a plate (chapter 10).
- Strong earthquakes are correlated with strains that build up in the rock as plates move about (chapter 10).
- The patterns of landforms that are produced by various plate interactions help provide evidence that plates are moving (chapter 10).

<u>10</u> Volcanoes and Earthquakes

Objectives

In this chapter you will learn that:

- Explosive volcanoes occur near plate boundaries where subduction is taking place.

- Shield volcanoes erupt without violent explosions at midocean ridges and hot spots.

- When a strain builds up in the crust that is beyond the ability of rocks to hold, the rocks yield, and the energy is released as an earthquake.

- The Richter scale is one measure used to rank the strength of earthquakes.

- Each whole-number increase on an earthquake scale like the Richter scale (e.g., 3.5 to 4.5 or 5 to 6) represents a tenfold increase in earthquake strength.

- A fault is the plane that rocks slip along to release strain.

Plates, Volcanoes, and Earthquakes

Volcanoes and earthquakes are geologic phenomena that are very closely related to plate tectonics. Volcanoes are formed when magma (molten rock) from deep within Earth reaches the top of the crust and escapes up to the ground (or seafloor). This is most likely to happen at plate edges because of the gaps that open when plates diverge, or because of subduction processes. There are two principal types of volcanoes:

1. the explosive kind (stratovolcanoes)

2. the other kind (shield volcanoes)

An earthquake is a shaking of the ground produced by the release of energy. Earthquakes can happen anywhere, but strong earthquakes are most likely to happen near plate boundaries because of the way that the movement of the plates causes a buildup of strain within Earth. Nearly all strong earthquakes happen near plate boundaries.

Where are there likely to be volcanic eruptions or strong earthquakes?

———————————

Answer: at plate boundaries.

Explosive Volcanoes

Explosive volcanoes occur near plate boundaries where subduction is taking place. As the subducting crust is forced down deep into Earth, it melts. That molten material is less dense than the surrounding rock, so it can work its way back up toward the surface. Apparently the magma acquires some of the properties of the rock it travels through as it moves toward the surface. The molten material in explosive volcanoes is relatively thick and viscous and traps a lot of gas—consequently, if the pressure builds up too high, a volcano will explode to release the pressure out to the atmosphere.

When the volcano explodes, molten rock is ejected out, onto the surface of Earth. Since it is molten, it will flow downward under the influence of gravity. When this lava cools, it will solidify into rock. As the lava cools and hardens, the ash and dust ejected into the atmosphere fall back to Earth, where they will land atop the newly solidified lava

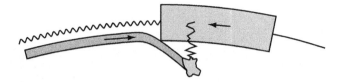

Figure 10.1. When there is collision and subduction at a plate boundary, magma from the molten subducting plate can work its way up, through the overlying crust.

rock. After many cycles of these explosions, a large, cone-shaped form of alternate ash and rock layers will be created. It is similar in shape to the form created when sand is dropped (e.g., onto the ground or in an hourglass). These layers are the reason why explosive volcanoes are also called *strato*volcanoes.

Explosive volcanoes are dangerous to people because:

- The explosion itself can be dangerous.

- Hot, flowing lava can be dangerous.

- The dust and ash can be a health hazard and also can settle back to Earth in large quantities.

- If the volcano is high and snow-covered, an explosion can melt the snow, causing floods or mudslides.

- Hot toxic or suffocating gases can be vented out of volcanoes.

Figure 10.2. *Left:* The landform created by an explosive volcano is a cone-shaped feature made up of alternating layers of rock and ash. *Center:* The rock comes from magma and lava, which flow down the volcano and solidify. The ash layer is deposited atop the new rock when material ejected into the atmosphere falls back down to Earth. *Right:* Each successive eruption deposits new layers of rock and ash, which build up the cone-shaped form.

The cone-shaped explosive volcano Mount Cleveland is in the Aleutian Islands. This 1,730-meter (5,676-foot) stratovolcano has been the site of numerous eruptions in the last two centuries, most recently in 1994. (Photograph by M. Harbin, University of Alaska–Fairbanks, from U.S. Geological Survey Digital Data Series DDS-40, version 1.1.)

What types of materials make up the layers in a stratovolcano?

Answer: One layer is rock formed from hardened lava. Another layer is dust, ash, and other remnants of the explosion that were ejected into the sky and fell back to Earth.

Shield Volcanoes

Shield volcanoes are volcanoes that do not erupt violently. The magma that comes out of shield volcanoes flows easily and therefore does not trap as much gas as the magma in an explosive volcano. When a shield volcano erupts, the magma flows out onto the solid part of Earth (under either air or water). This magma flows very easily and will eventually solidify into new rock.

Shield volcanoes usually are found in connection with oceanic crust:

• They are found at midocean ridges where the splitting apart of two plates weakens the crust and allows molten material to come out of the solid Earth.

• They are found at hot spots in the sea.

A hot spot is a place where volcanic activity is found in the middle of a plate—not at the edge of a plate. The Hawaiian Islands are an example of a hot spot. The islands were formed by the eruption of shield volcanoes at an uncharacteristic location in the middle of the Pacific Plate. Shield volcanoes can be dangerous to people because the extremely hot, easily flowing lava can travel great distances before it cools and solidifies.

What is a "hot spot"? _____

Answer: It is a place where volcanoes are found uncharacteristically in the middle of a plate instead of near a plate boundary, where they are typically located.

Earthquakes

When strain accumulates within Earth's crust (e.g., from plates moving around), it can build up until the rocks reach the breaking point. If strain reaches the breaking point, the rocks slip and the strain is released into the environment as energy.

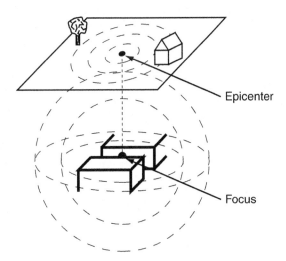

Figure 10.3. If strain builds up beyond the capacity of rock to absorb it, the rock will yield—slip—to relieve the load. The slippage frees the built-up strain from the rock and releases it into the environment as earthquake energy. The point where the slippage occurs is called the earthquake's focus. The point directly above the focus on the surface of Earth is the earthquake's epicenter.

The place within Earth where the slippage occurs is the earthquake's *focus*. The place on Earth's surface that is directly above the focus is the earthquake's *epicenter*. The epicenter is the place on Earth's surface that is closest to the focus. Waves of energy radiate away from the focus—away from the place where the strain is released. When earthquake waves reach the surface, their energy can cause shaking or movement. This can rearrange a landscape or cause damage to human structures.

How does an earthquake cause shaking? _____

Answer: When accumulated strain is too much, the rocks will slip and release the energy into the environment.

Measuring Earthquakes

The more strain that is released by slippage at the focus, the greater the shaking that can occur. The amount of damage from an earthquake is highly correlated with the amount of energy released.

The Richter scale is one measure of how much shaking takes place. Although there are a few scales used to report earthquake intensity, they have all been calibrated to give similar ratings. Earthquakes with magnitude 1, 2, and 3 are very common and usually don't do much damage—some aren't even noticed. The most damaging earthquakes are usually measured at 6, 7, and 8. The largest earthquake recorded was a magnitude 9.5. Because earthquake measurements are based on powers of 10, each increase of 1 whole unit on a scale means a ten times increase to intensity. A magnitude 4 earthquake is ten times more powerful than a magnitude 3. A magnitude 7 earthquake is a hundred times more powerful than a magnitude 5.

If a Richter scale 6 earthquake happens in Japan and a Richter scale 4 earthquake happens in California, how many times more powerful is the Japanese earthquake? _____

Answer: It is a hundred times more powerful ($10^6 \div 10^4 = 10^2 = 100$).

Faults

The plane along which the rocks slip to release the strain and cause an earthquake is called a fault. There are four types of faults that are produced by the forces that build up strain in the rock:

Richter Scale Intensity

A Richter scale 3 earthquake has ten times more shaking than a Richter scale 2 earthquake has.

$10^3 \div 10^2 = 10^1$, which is 10

Not $3 - 2 = 1\times$ as powerful (Obviously, right? This is how you know that this is not the right way to do this calculation.)

Not $3 \div 2 = 1.5\times$ as powerful

It is $\dfrac{10^3}{10^2} = \dfrac{1,000}{100} = 10\times$ as powerful.

A Richter scale 6 earthquake has 10,000 times more shaking than a Richter scale 2 earthquake has.

$10^6 \div 10^2 = 10^4$, which is 10,000

Not $6 - 2 = 4\times$ more powerful

Not $6 \div 2 = 3\times$ more powerful

It is $\dfrac{10^6}{10^2} = \dfrac{1,000,000}{100} = 10,000\times$ as powerful.

A Richter scale 7.5 earthquake has 100,000 times more shaking than a Richter scale 2.5 earthquake has.

$10^{7.5} \div 10^{2.5} = 10^5$, which is 100,000

Not $7.5 - 2.5 = 5\times$ more powerful

Not $7.5 \div 2.5 = 3\times$ more powerful

It is $\dfrac{10^{7.5}}{10^{2.5}} = \dfrac{31622777}{316.22777} = 100,000\times$ as powerful.

1.   normal fault

2.   reverse fault

3.   overthrust fault

4.   strike–slip fault

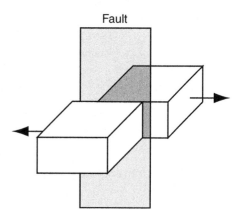

Fault

Figure 10.4. A fault is the plane along which rocks slip to release accumulated strain.

Normal Fault

If the rock is affected by forces that put it into tension (i.e., it is being pulled apart), the slippage will produce a normal fault. In normal faulting (depending on the direction of the tension forces and the way the rock breaks), one section of Earth will be raised (or lowered) with respect to the other section.

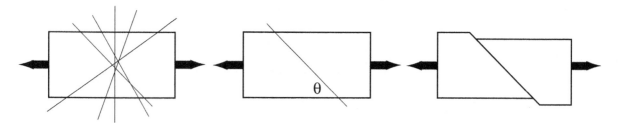

Figure 10.5. *Left:* A rock that is being pulled apart can break in an infinite number of ways. It doesn't matter which path it breaks on; all the potential paths shown would give a similar result. *Center:* If the tension builds up beyond the capacity of the rock to bear it, the rock will break along some angle. *Right:* The rock will yield and slip along the fault.

Reverse Fault

If the rock is affected by forces that put it into compression, the slippage produces a reverse fault. Compression also will raise or lower one section of Earth with respect to the section on the other side of the reverse fault.

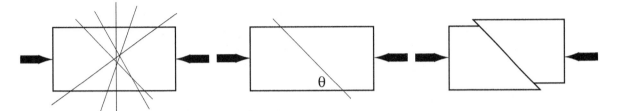

Figure 10.6. *Left:* A section of Earth's interior that is being compressed by tectonic forces could break along any of numerous possible paths. *Center:* If the compression force is too much for the rock to sustain, then it will yield—slip— and release the built-up energy. *Right:* In a reverse fault, compression will raise or lower one section of Earth with respect to the other.

Overthrust Fault

If compression continues along a reverse fault, the reverse fault can be transformed into an overthrust fault. In an overthrust fault, one section of rock is pushed over the other section.

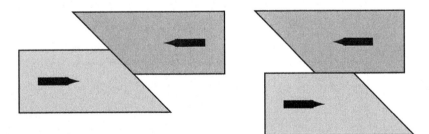

Figure 10.7. *Left:* As compression continues in a reverse fault, one segment of rock is lifted up relative to the other. *Right:* Sometimes compressive force can lead to one rock section riding over the adjacent section. When a section of rock is pushed past another and begins to slip over the top, this is called an overthrust fault.

Strike-Slip Fault

The fourth way in which two sections of crust can interact is known by a variety of names: strike-slip fault, transverse fault, or transcurrent fault. These refer to situations where one section of Earth's crust slides past the

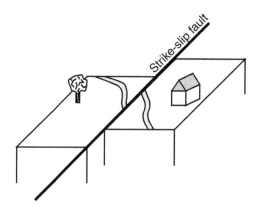
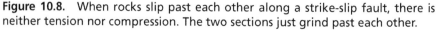

Figure 10.8. When rocks slip past each other along a strike-slip fault, there is neither tension nor compression. The two sections just grind past each other.

other without pushing toward or pulling from the adjacent section. Strain can build up along the boundary, but it is a shearing force, not compression or tension. If the rocks yield to the force, then the strain that has built up will be released into the environment as earthquake energy.

When rocks are stressed and strained, they will yield along paths that are determined by the nature of the applied forces (tension, compression, shearing) and by the internal qualities and strength of the rock itself. The rock can yield in three dimensions, not just in the ways shown in this book's two-dimensional diagrams. Thus strike-slips can be horizontal, vertical, or some combination, and they also might occur along with tension or compression.

The lateral offset in this plowed field is a product of movement along the Imperial fault during a 1979 California earthquake. (Image courtesy of National Geophysical Data Center.)

How can an earthquake uplift a section of Earth? _____

> *Answer:* In reverse or overthrust faulting, the compression forces can push a section upward. In normal faulting, a section can be pulled upward.

Evidence for the Theory of Plate Tectonics

The theory of plate tectonics holds that the surface of Earth is made up of individual plates that can move independently. The movement of these plates carries the landmasses and oceans along. Interactions of these plates with their neighbors produce certain landforms. Earthquakes and volcanoes are particularly associated with plate tectonics. Subduction and plate divergence are strongly correlated with volcanoes. And the grinding of neighboring plates is a precursor to strong earthquakes.

Plate tectonics and its geomorphic consequences have been discussed in this chapter and in chapters 8 and 9. There is a weight of evidence that makes plate tectonics—the idea that separate pieces of the lithosphere can move independently across Earth—the best explanation for how Earth has physically come to be and how it continues to evolve geographically.

Some of the evidence for the theory of plate tectonics:

- There are midocean ridges in the center of oceans.

- There are systematic mountain chains, not a random distribution of mountains.

- Worldwide pattern of earthquakes is *systematic,* not random.

- Worldwide pattern of volcanoes is *systematic,* not random.

- Deep ocean trenches are systematically located just offshore from landmasses or parallel to island chains.

- Stratovolcanoes and shield volcanoes are found in different places.

- Age of sea floor rocks increases with distance from midocean ridges.

- African Rift valleys and the Red Sea show that a continent can split apart.

- Oceanic islands are found in chains.

- Continent shapes appear to fit together very well across ocean gaps.

Other evidence, not described in this book:

- Patterns of paleomagnetism pole reversals indicate seafloor spreading.

- Magnetic directional evidence of polar wandering shows that plates move.

- All oceanic crust is younger than all continental crust.

- Seafloor sediment gets thicker with distance from midocean ridges.

- Hot-spot traces show where plates moved.

- Folding and warping of rock layers show that large-scale forces are applied to landmasses.

- Similar plant and animal fossils are found on now widely separated landmasses.

- There have been paleoclimates that differed from current climate regimes.

- Accreted terranes indicate that plates are moving.

- Subduction is supported by earthquake positions and depths near trenches.

SELF-TEST

1. A Richter scale 8 earthquake is how many times more powerful than a Richter scale 4 earthquake?
   a. 2   c. 32
   b. 4   d. 10,000

2. Which type of faulting occurs when tension forces within Earth are pulling rock apart?
   a. reverse fault   c. overthrust fault
   b. normal fault   d. strike-slip fault

3. An explosive volcano is associated with _____.
   a. hot spots   c. subduction
   b. normal faulting   d. Hawaiian Islands

4. The _____ is on the surface of Earth; it is directly above an earthquake's focus.

5. A _____ type volcano is found at midocean ridges.

6. Nearly all strong earthquakes happen in the middle of plates. (True or False)

7. Reverse faulting can lift up a section of Earth relative to an adjacent section. (True or False)

8. How does the magma in an explosive volcano differ from the magma in a shield volcano?

9. Why is an explosive volcano sometimes called a stratovolcano?

10. What happens to the molten rock known as magma to transform it into the molten rock known as lava?

ANSWERS

1. d   2. b   3. c   4. epicenter

5. shield-   6. False   7. True

8. Magma in an explosive volcano is more viscous and holds more compressed gases.

9. Because it is comprised of alternating layers of solidified lava and ash fallout that were released during past explosions.

10. Magma is molten rock within the Earth. Lava is magma that has escaped to the solid surface of Earth (which may be underwater).

Links to Other Chapters

- Ash from volcanic eruptions becomes one of the constituents of the atmosphere (chapter 3).
- Volcanic ash in the atmosphere can intercept insolation (chapter 3) and affect global climates (chapter 7).
- Explosive volcanoes are associated with subduction (chapter 9).
- Shield volcanoes are associated with plates splitting apart (chapter 9).
- Solid rock can be created within Earth if magma solidifies before it reaches the surface (chapter 11).
- Deposits of volcanic ash provide mineral nutrients that can aid plant growth (chapter 11).

11 Weathering

Objectives

In this chapter you will learn that:

- Weathering is the process that breaks rock into smaller, more easily transported pieces.
- Physical weathering processes break a rock into smaller pieces, each of which has the same properties as the original piece.
- Chemical weathering breaks down a rock through chemical reactions, which transform the rock into some other kind of material.
- Water is a critical element for weathering.
- Gravity acts to pull all material down, toward the center of Earth.
- Soils are created when weathering processes act on rock near Earth's surface.
- Soils usually develop a sequence of layers called horizons.
- Climate is an important control on how a soil develops.

Denudation

If a rock is near the surface of Earth, it will be attacked by agents of denudation. Denudation is the process of lowering the surface of Earth, and it is accomplished by weathering, which breaks rock into smaller pieces, and by the actions of water, air, ice, and gravity, which bring the rock particles from higher places to lower places. The breaking of rock into smaller, more erodible pieces is important. It is much easier for water, air, ice, and gravity to move many small pieces than one large rock.

There are two types of weathering: physical weathering and chemical weathering.

Why does it matter that weathering breaks rock into smaller pieces?

Answer: Smaller pieces are more easily moved than larger pieces are.

Physical Weathering

Physical weathering (sometimes called mechanical weathering) breaks rock into smaller pieces of itself. The little pieces have the same properties as the original bigger piece. It would be theoretically possible to reconstruct the original rock to its original size and shape if all of the physically weathered pieces were at hand.

There are many ways to physically weather rock, but all either shatter, chip, split, or burst a rock to break it into smaller pieces. Rock can be weathered into smaller pieces by contact forces, ice, plant growth, salt crystallization, unloading, and other physical processes, as described below:

Contact Forces

Physical weathering happens if a force applied to a rock breaks it into smaller pieces. A rock falling onto a hard surface and shattering, or sand carried by water or wind that abrades other rocks are examples of contact forces that weather rock.

Ice

Ice is another physical weathering agent. When water freezes, its volume increases (it takes up more space). This is why a can of soda will

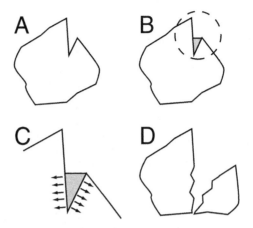

Figure 11.1. In this sequence from A to D, water gets in a crack in a rock and freezes. The expanding ice puts pressure on the rock and breaks it.

explode if it is left in the freezer; it is why an ice cube tray will overflow with ice although it was only topped off with water. If water gets into a rock (e.g., into a crack) and then freezes, the expanding ice will put a force on the rock. Repeated cycles of freezing and thawing can enlarge a crack or break off a piece of the rock.

Plant Growth

Plants can similarly break apart a rock. If a plant grows in a crack, or if it can extend a root into a crack, as the plant or root grows larger, its expansion will put a force on the rock. There may be enough force to split the rock into smaller pieces, or weaken the rock, or widen the crack.

Figure 11.2. If plants grow in a rock, or if a plant's roots get into a crack in a rock, as the plant or root grows larger, it will apply a splitting force to the rock.

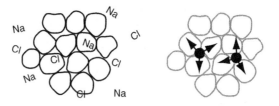

Figure 11.3. *Left:* If water with dissolved salt can get into a porous rock such as sandstone or concrete, it can bring salt ions with it into the rock matrix. *Right:* When the water evaporates, the recrystallizing salt will pop out of solution inside the rock. Its sudden appearance within the rock can cause a splitting force that might loosen or dislodge a particle.

Salt Crystallization

When salt comes in contact with water, the salt dissolves. The salt breaks apart chemically and becomes part of the liquid solution. If, later, the water evaporates, the salt particles will recrystallize back into solid pieces. You may have seen such salt stains on a cooking pot; salt also may form on your forehead or brow as your perspiration dries.

Salt crystallization is another physical weathering process. Water with dissolved salt can get into porous rocks like sandstone. When the water dries up, the salt will recrystallize. When salt crystals materialize inside of a rock, they can put a force on the rock that can help break it apart.

Unloading

Unloading is another physical weathering process that helps break rock apart. Rock can form within Earth instead of on top of it. Sometimes magma might rise from a lower level but not reach the surface. It will cool and solidify under pressure within the solid Earth. Sedimentary or metamorphic rock also is created when pressure from overlying layers acts to transform Earth material into solid rock.

If rocks are created under pressure, and some of the overlying material is eroded away over time, this will relieve some of the force on the rock. As outside pressure is released, the rock will relax, or rebound slightly. This could cause the rock to break.

Cracking or exfoliating can either cause pieces to break off, or allow water, ice, or plant roots in. So, unloading (release of confining pres-

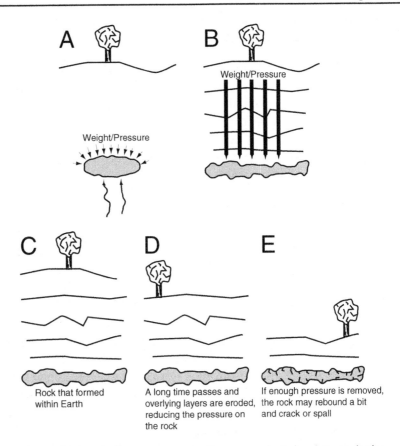

Figure 11.4. *A:* If magma works its way up but doesn't reach the surface, it will cool and solidify within Earth under pressure from overlying material. *B:* Sometimes great pressure from overlying layers can transform loose Earth materials into rock. *C:* A rock formed under pressure is "used to" that pressure. *D* and *E:* If erosion of the ground releases enough of the pressure, that might allow the rock to relax and expand a bit. This can cause it to crack or exfoliate.

sure) is another physical weathering process that contributes to the breakdown of rock into smaller pieces.

Why is salt crystallization a physical weathering process?

Answer: The recrystallization of salt inside a rock puts an internal force on the rock that can break it into smaller pieces. The pieces have the same properties as the original material.

Chemical Weathering

Chemical weathering helps reduce a rock to smaller pieces through chemical reactions. The products of chemical weathering are different from the original rock material. You could not put the rock back together to its original form even if you had all of its chemically weathered pieces at hand. Oxidation and carbonic acid action are two common chemical weathering processes whose effects are easily noticed, as described below:

Oxidation

With oxidation, the minerals that make up the rock react with oxygen from the atmosphere, or oxygen dissolved in water. Iron is the preeminent example of a material susceptible to oxidation. When iron reacts with oxygen, it forms rust (iron oxide, Fe_2O_3). Rust does not have the same properties as iron. Rust, for example, does not have the strength of iron. It is much easier to erode rust; you can fleck rust away with your finger. When shiny metallic copper turns bluish-green, that is another example of oxidation. A substance is altered when it reacts with oxygen and is subsequently more easily eroded than it originally would have been.

Carbonic Acid

Acids also can cause chemical weathering. When carbon dioxide dissolves in water, it forms carbonic acid. Carbon dioxide gas gets in liquid water the same way that the oxygen that fish breathe gets into water. Soda you buy in a grocery store is acidic because the introduced carbonation forms an acid in the drinking solution, although in nature carbonic acid solutions are much weaker than the soda we buy.

Carbonic acid can help dissolve rocks. In particular, limestone rock is very susceptible to carbonic acid action when acidified water is in contact with it. Limestone is easily dissolved (broken down) by the acid. As the water flows away, the dissolved limestone material is carried off, too.

How is carbonic acid created in nature? _____

> *Answer:* Carbonic acid is created when carbon dioxide becomes dissolved in water.

The grave marker at top is dated 1804, and the words are still clearly inscribed in the resistant stone. In contrast, the limestone grave marker from 1812 *(bottom)* is difficult to read because acid weathering has blurred the lettering. (Photographs courtesy of MACGES.)

Water

Water is a critical agent in the weathering process. Moving water can carry particles that abrade rock into smaller pieces. Water carries dissolved salt particles with it as it seeps into a rock's matrix. Water within a rock could freeze and apply a splitting pressure. Dissolved oxygen in water can promote rust. Dissolved carbon dioxide in water creates carbonic acid. The presence of water is necessary for plant growth, and plants also can act to weather rock.

As water flows downward, it can mobilize and transport sedimentary particles. Rock pieces that are abraded in a stream or dissolved by carbonic acid will find themselves immediately transported by moving water. If a flow of water preceded freezing, ice wedging, salt crystallizing, or oxidizing, then, if that flow resumes, the water might be able to mobilize and transport the newly weathered smaller rock particles.

Because weathering is so closely linked with liquid water, climate is an important factor in weathering rates. Water can only dissolve and seep and flow in places where it is available. In general, when water is available, warm temperatures promote chemical reactions, evaporation, and plant growth, while cold temperatures lead to freeze-thaw cycles.

How does temperature influence the weathering of rock?

Answer: Temperature affects the rate of chemical weathering and evaporation. It also can lead to plant growth or ice.

Gravity

The importance of gravity in denudation cannot be overstated. Gravity is the force that pulls everything toward the center of Earth. (Toward the center of Earth is synonymous with "down".) Gravity can work directly on an object to pull it downward. Gravity also can pull down on water or ice, which then may move rocks or sediment with it.

The steeper the slope, the easier it will be for gravity to move something—anything—downward: rocks, water, ice, bicycles. As a slope nears vertical, gravity can act more and more directly. On a flat, hori-

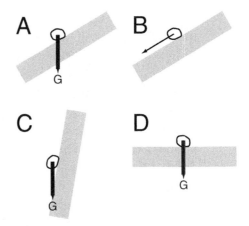

Figure 11.5. *A:* A rock on a slope is pulled downward by gravity—toward the center of Earth. *B:* As a result of being pulled toward the center of Earth, it will move (slide or roll) along the slope, which is the only path downward it can travel in response to gravity's pull. *C:* As a slope nears vertical, gravity can act more and more directly. The path the object travels becomes more similar to the direction of gravity's attractive force. *D:* If an object is on a flat surface, then gravity will not be able to move it downward (although it still will pull on it).

zontal surface, gravity cannot move an object because the object is unable to get closer to the center of Earth.

What role does gravity play in denudation? _____

> *Answer:* Gravity acts to bring particles to lower elevations. It can either act directly on an object, or it can cause water or ice to move downhill and transport material with it.

Rock Weathering to Soil

Weathering and groundwater (see chapter 12) are two principal actors that, working together, can create soil. Soil is not just pieces of rock, although those pieces are crucial parts of soil. Soil can be viewed as a system in which weathered rock provides an environment in which plants can absorb nutrients and grow.

Soil formation begins when bedrock becomes exposed to weathering forces. This does not always mean right at the surface—weathering can be effective belowground. As weathering agents begin to work on the rock, they will start to break it down into smaller pieces. After some time, weathering will establish a three–layer system in the ground. The bottommost section will be solid rock that is too deep to be affected by weathering processes. Above that will be a section at the limits of weathering, where the solid rock is beginning to break apart. At the top, the rock will have been exposed to stronger weathering for a longer time, and the mineral pieces will be small and more broken down.

As the top layers of the rock are broken down into smaller pieces, weathering agents can penetrate deeper into the rock. Once the top layer of broken pieces is established, an opportunity is provided for plants to colonize the area because plants can pull mineral nutrients from the small weathered rock pieces. When plants initially establish themselves in these highly weathered rock particles, that can lead to the development of a true soil. Animals will follow the plants, and the further action of animals and plants will help to break up the ground and provide a path to extend weathering processes downward.

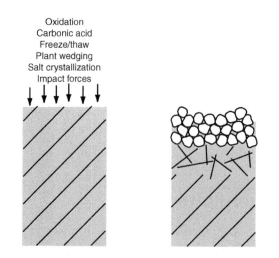

Oxidation
Carbonic acid
Freeze/thaw
Plant wedging
Salt crystallization
Impact forces

Figure 11.6. *Left:* As various weathering forces begin to attack solid rock, they will produce a three-zone sequence in the rock. *Right:* There will be a section of rock that is too deep to be affected by weathering. Above that will be a zone at the limits of weathering, where rock is beginning to break apart. At the top, weathering forces will have been working for some time, and the rock will be broken into small, individual pieces.

What is soil? _____

> *Answer:* Soil is a physical system in which weathered rock can support the growth of plants.

Soil Horizons

The initial three-layer system of solid rock at the bottom, well-weathered smaller particles at the top, and rocks at an intermediate stage between them, will be further modified as organic material is introduced and the downward movement of water begins to transport material within the ground. The layers within the ground are known as horizons.

A soil is made up of weathered rock and organic material. It can be found in a layered arrangement that is produced by weathering forces and the movement of water within the ground. The layers are the O- (organic) horizon at the top; unweathered parent material—bedrock—is the bottom limit; in between, the horizons run A–B–C, from top to bottom.

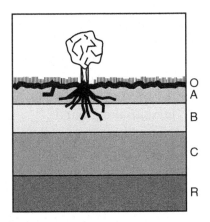

Figure 11.7. Weathering, biological processes, and the movement of groundwater can work together to produce a soil that is made up of discrete horizons. Solid rock (R) that is too deep to be weathered and a layer comprised solely of organic material (O) at the top bound the soil. The top mineral layer (A) has much organic material within it. When well-weathered, fine-grained material is washed downward out of the A-horizon, it accumulates to produce a new layer, B. The C-horizon is comprised of rock at the limit of weathering effectiveness that is beginning to be broken apart.

C-Horizon

It has been established that at some depth the bedrock will be too far belowground to be affected by weathering processes and thus will remain as solid rock (sometimes called the R–horizon). However, above that level, at the limits of weathering effectiveness, there will be a zone called the C-horizon, where solid rock is beginning to break apart. The rock pieces there are still easily recognized as being the same material as the solid rock below them.

O-Horizon

As rock weathering continues, plants will die and decay, leaves and twigs will drop to the ground, animals will die and decay. The very top of the ground will now be made of a layer of organic material (leaves, stems, twigs, decaying biota) that is largely mineral-free. This will be a new soil layer, the O- (organic) horizon.

A-Horizon

In the middle (between the top O-horizon and the partially weathered rock of the C-horizon) will be a two-layer system of weathered mineral matter. The upper layer, the A-horizon, will have the most weathered material and also will have a significant amount of organics. Some of the organic material was directly emplaced in the layer by animals or plant roots. Some of the organic material was washed into the layer as decaying material in the O-horizon was brought down by water seeping into the ground.

B-Horizon

As the top mineral layer gets more and more weathered, and its particles get smaller and smaller, they also can be washed downward by moving water. This fine-grained mineral matter will be washed out of the A-horizon, and its accumulation at a lower level will create a new soil layer, the B-horizon.

Why is the A-horizon, which is sometimes called topsoil, so important for the growth of plants? _____

> *Answer:* Plants can access nutrients from the small, well-weathered mineral pieces, and from decaying organic matter that is in this soil layer.

Soil-Forming Factors

Not every place will have the typical O–ABC–R profile of a well-developed soil. The formation of a soil is a complex phenomenon that depends on many factors working together. Different places will have different kinds of soil, depending on the following:

- The type of rock that is weathered will affect the mineral and chemical composition of the soil and its ability to nourish plants.

- Climate will affect the kinds and rates of weathering processes that work on rock. Climate also affects the quantity and type of plant and animal life that will be present.

- The type of plant and animals in the environment affects soil. Some plants and animals will accelerate weathering. The death and decay of organisms provide nutrients for further plant growth.

- Topography, especially slope, will affect what proportion of precipitation will be runoff, which can erode soil material, or infiltration, which aids soil weathering. Infiltration brings water into the ground and thus introduces some weathering processes to the subsurface.

A soil profile was revealed when a foundation was excavated for a house. This is a more complex profile than the model discussed in this chapter. However, you can see the uppermost mineral layer—the dark A-horizon—and the organic grass just above it. Neither the C-horizon nor a bedrock layer is exposed here. (Photograph courtesy of MACGES.)

• It can be a very long time before processes can create a well-developed soil. In contrast, soil can be eroded away very quickly.

How does climate influence the development of a soil? _____

Answer: Climate is a control on the type of weathering processes that are present, on the quantity of water that can move within the ground, and on the types of plants and animals that will be present.

SELF-TEST

1. What physical weathering process erodes rock in a moving stream?
   a. contact forces
   b. unloading
   c. ice wedging
   d. salt crystallization

2. Which of these is most likely to be an active weathering process when underground water is in contact with limestone rock?
   a. unloading
   b. contact forces
   c. carbonic acid action
   d. salt crystallization

3. Which type of place is most likely to have well-developed soils?
   a. cold, wet, steep
   b. cold, dry, flat
   c. warm, dry, steep
   d. warm, wet, flat

4. What always operating force can cause water or ice to move rock particles down to lower elevations?

5. A common oxidation process occurs when iron is transformed into more easily erodible pieces called _____.

6. If a plant produces an acid that helps break down a rock it is growing in, that would be a chemical weathering process. (True or False)

7. If you had all of the pieces from a chemically weathered rock, it would be theoretically possible to fit them back together and reconstruct the original rock. (True or False)

8. How does a place's temperature affect the types of weathering that are effective there?

9. How does weathering contribute to erosion (the mobilization and transport of material)?

10. How does organic material get into a soil's A-horizon?

ANSWERS

1. a   2. c   3. d   4. gravity

5. rust   6. True   7. False

8. Cold places can have ice processes. Warmer places have more evaporation, have greater plant activity, and have faster-paced chemical reactions.

9. When weathering breaks rock into smaller pieces, transporting processes can move them more easily.

10. Some can be emplaced directly from plants or animals that live in the layer. Some can be washed into the layer by water seeping down from the surface.

Links to Other Chapters

- For ice to be an effective weathering agent there needs to be a cold climate (chapter 7).
- Salt crystallization is most effective in hot, dry places with a lot of evaporation (chapter 7).
- Water is a critical component in much of chemical weathering, so a wet climate can aid chemical weathering (chapter 7).
- The evolution of soil horizons, especially A and B, require ground-water movement (chapter 12).
- Unweathered bedrock can be a barrier to the movement of ground-water (chapter 12).
- Weathering and the size of soil particles affect the rate of infiltration (chapter 12).
- Gravity brings the water in overland flow and streams to lower elevations (chapter 13).
- Rocks must be broken down to small sizes before the wind can effectively transport them (chapter 14), and windblown sand can be a physical weathering agent.
- Gravity will bring saltating particles back to the ground (chapter 14).
- Gravity causes glacial ice to move to lower elevations (chapter 14).
- Waves breaking on a shoreline can physically weather rock (chapter 15).

12 Groundwater

Objectives

In this chapter you will learn that:

- When water is at the surface of Earth, it must either run off across it, or infiltrate into the ground.

- Water that infiltrates becomes part of the groundwater system.

- Ground that is completely filled with infiltrated water is saturated.

- The ground below the water table is saturated, while above the water table there is free space.

- A well is a hole that reaches below the water table into saturated ground so that water can be removed from the ground for human use.

- If the water table intersects the ground surface, water will flow out onto the surface.

Water at Earth's Surface

Rain, dew, or snow—any of these (and many other) events—will cause water to end up on the surface of Earth. Once water is on the solid surface of Earth, it *must* do one of two things:

1. infiltrate—seep into the ground

2. run off—move across the surface to a lower elevation

Water also can evaporate, but if it doesn't evaporate instantaneously upon its arrival at the surface, then it must infiltrate or run off first.

Infiltration and runoff are each the result of gravity working to pull water drops closer to the center of Earth. This chapter will address infiltration and groundwater (water inside Earth). The following chapter will address runoff (water moving across the surface of Earth).

What are the two things that a drop of water can do if it reaches the solid surface of Earth? _____

Answer: It can infiltrate (seep into the ground) or run off (move across the surface).

Infiltration

To infiltrate, individual water drops must move between the rocks and rock particles to get below the surface. Several factors can influence the proportion of water that will infiltrate or run off:

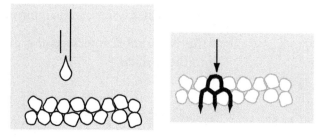

Figure 12.1.   Water infiltrates if it can seep between soil particles and get below the surface of Earth.

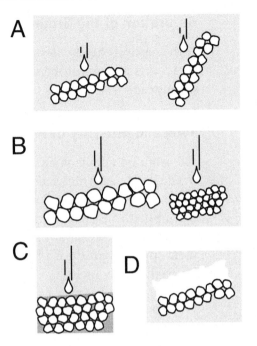

Figure 12.2. *A:* More water will infiltrate on flatter slopes than on steeper slopes. *B:* It is easier for water to infiltrate if ground particles are large and there are bigger spaces between them. *C:* If the ground is already saturated, there is no room for more water, and there can be no infiltration. *D:* Slowly melting snow can produce a lot of infiltration, whereas other types of precipitation events might end up with nearly all runoff.

Slope of the Ground

Water that falls on flatter ground will be more likely to infiltrate than water that falls on steeper slopes.

Size of the Ground Particles

Water falling on bigger particles (with larger spaces between them) is more likely to infiltrate than water falling on little particles (with small spaces between them). Coarse sand and gravels have more infiltration because the space between the grains is large enough for water to easily pass through. Clay is comprised of very small particles and allows almost no water to pass.

Saturation of the Ground

If the ground already has water filling all the space between its particles, new water arriving at the surface will be unable to infiltrate and will run off instead.

Type and Intensity of Precipitation

A lot of rain falling all at once will promote runoff because only some of the newly arriving water will be able to get quickly into the ground. In contrast, the same amount of rain drizzling over several days might end up with nearly complete infiltration. If snow falls and very slowly melts, then nearly all of the water can infiltrate. If a lot of snow rapidly melts, that can produce a large amount of runoff (especially if it is raining as the snow melts).

Why does water infiltrate easily into gravel or sand, but hardly at all into clay soils? _____

> *Answer:* Gravel and sand are relatively coarse and have larger spaces between the particles that water can flow through. Clay is very fine-grained, and to a drop of water, the gap between particles is impossibly small to pass through.

Groundwater Movement

Gravity still will work on water, even after it infiltrates into the ground. Gravity pulls water very slowly through the ground. The water has to make its way through the spaces between all of the particles that make up the ground. Nevertheless, gravity always will be working to pull the water closer to the center of Earth. Sometimes there is an obstacle in the way of

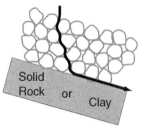

Figure 12.3. Groundwater is pulled down slowly by gravity. Solid rock or a clay layer can stop or redirect the downward movement of water particles.

the water's downward path. If the obstacle is an impenetrable barrier—solid rock, or a clay layer—then the water cannot move through it.

What causes water to move through the ground after it infiltrates?

Answer: Gravity always is pulling water drops toward the center of Earth, and this will cause the water to move through the ground.

Saturation and the Water Table

The ground is said to be saturated if all the spaces between the ground particles are filled with water. The ground is holding as much water as it possibly can. There is an imaginary line that separates the saturated ground below the line from the unsaturated ground above that line. This line is called the water table.

If a hole is dug into the ground and the hole goes deep enough to get below the water table, then the hole will fill with water. The ground below the water table is saturated—all the space between the ground particles is filled with water. It doesn't matter how big the space is.

Sometimes holes are purposely dug below the water table so they will fill with water. A well is simply a hole in the ground that reaches below the water table—and then fills up. Water can be withdrawn by pumping or scooping it out of the hole.

The water table marks the boundary between saturated and unsaturated ground. However, the position of the water table is not fixed. It can go up or down depending on how much water is being added or

Figure 12.4. The water table separates saturated ground _(below)_ from ground that has free storage space _(above)._

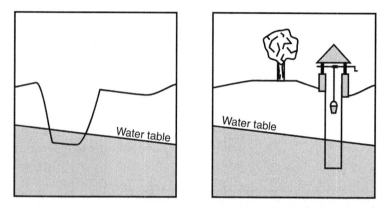

Figure 12.5. *Left:* A hole dug below the water table will fill with water. All the space below the water table is saturated—it doesn't matter how big the space is. *Right:* A well is a hole that purposely reaches below the water table so it will fill with groundwater for easier withdrawal.

removed from the ground. If there is a net increase in the amount of water in the ground, then the water table will rise as more and more of the ground becomes saturated. Infiltrating water from rain, flooding, snow melt, or irrigation increases the amount of groundwater and raises the water table. In contrast, there can be a net decrease in the amount of groundwater if there is a decrease in infiltration. Infiltration can be reduced by drought. A change to the ground surface that acts to

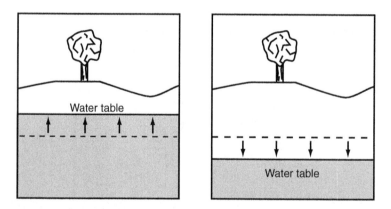

Figure 12.6. *Left:* Infiltrating water from rain, snow melt, or floods will add water to the ground and raise the water table. *Right:* A decrease in infiltration will reduce the amount of water in the ground and lead to a lower water table.

Figure 12.7. If the water table drops below the bottom of a well, then the well will be in the unsaturated part of the ground—and thus will be dry.

increase the amount of runoff instead of infiltration (e.g. building, paving, storm sewers) also will act to lower the water table. The water table also will be lowered if water is withdrawn from the ground at a well. If there is a drought or a very high rate of water withdrawal, the water table can drop below the level of a well, causing it to go dry.

How does a change in the amount of infiltration affect the depth of the water table? _____

> *Answer:* If infiltration increases, there will be more water in the ground, and the water table will rise. If infiltration decreases, there will be less groundwater, and the water table will drop.

Groundwater at the Surface

A spring is a place where the water within Earth comes out of the ground as a surface flow. Springs are found where the water table intersects the ground surface. Water released from springs is like all water at the surface of Earth. It will run off across the ground under the influence of gravity. Spring water can reinfiltrate into the ground, but it probably has to move away first, because the ground is likely to be saturated at the spring itself.

It is common for groundwater to fill up a topographic depression to create a spring-fed pond or lake. Where there is a groundwater-derived water body, the surface of the pond or lake is at the same elevation as the water table. Springs often are sources of water for streams, or they may provide additional water to flowing streams.

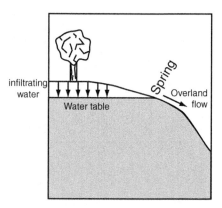

Figure 12.8. If the groundwater table intersects the surface of Earth, the water will flow out from the ground and move across the surface as runoff.

Many springs and ponds are seasonal phenomena that come and go over the course of a year. As the groundwater table moves up during a wet period and intercepts the surface, a spring may activate or a lake may fill. If the water table moves down in a dry season or a drought, then a spring may stop flowing and that source of surface water will dry up.

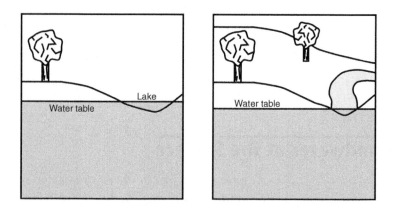

Figure 12.9. Groundwater can provide water to existing water features. *Left:* Groundwater can provide water to fill a lake or a pond. The elevation of the surface of a standing body of water is equal to the elevation of the water table. *Right:* Groundwater can flow into a stream, and it will then join that flow of water moving across the surface.

What is a spring? _____

Answer: A spring is a surface flow of water created when the water table intersects the surface of Earth and groundwater flows out.

SELF-TEST

1. When water moves between ground particles and seeps into the ground it is called _____.
   a. runoff
   b. infiltration
   c. withdrawal
   d. erosion

2. If all the empty space between ground particles is filled with water, the ground is said to be _____.
   a. drizzled
   b. melted
   c. irrigated
   d. saturated

3. Groundwater moves very slowly downward under the influence of _____.

4. A hole that is purposely dug below the water table to give people easy access to the groundwater is called a _____.

5. Rain falling on flat, coarse, gravelly terrain is more likely to seep into the ground than rainwater falling on a clay hillside. (True or False)

6. The line that separates saturated ground below from ground with free storage space above is called the water table. (True or False)

7. If there is a heavy rain that rapidly melts frozen snow, almost all of the water will seep into the ground. (True or False)

8. Why will a hole dug below the water table fill with water?

9. What happens to the water table if more water is withdrawn from the ground by people than is replaced by water seeping into the ground?

10. What will happen to the local water table if there is a drought?

ANSWERS

1. b   2. d   3. gravity   4. well

5. True   6. True   7. False

8. Any empty space (regardless of its size) that is below the water table will be filled with water.

9. If more water is taken from the ground than is replaced, the water table will be lowered.

10. As the amount of precipitation decreases, the amount of water that can infiltrate will decrease, and the water table will drop.

Links to Other Chapters

- The amount of groundwater in a place will depend on climate—how much water is added and how much water is evaporated away (chapter 7).
- Infiltrated water can freeze and lead to ice weathering (chapter 11).
- Carbonic acid in groundwater is a chemical weathering agent (chapter 11).
- Groundwater movement is important for soil formation (chapter 11).
- Groundwater is the supply that plants collect with their roots, and plant growth can be a weathering force (chapter 11).
- Groundwater can provide a base flow of water for streams (chapter 13).
- Groundwater that escapes to the surface at springs becomes surface water that runs off across the surface of Earth (chapter 13).

$\underline{\textbf{13}}$ Streams

Objectives

In this chapter you will learn that:

- All water on the surface of Earth that does not seep into the ground must flow downhill under the influence of gravity.

- Water flowing overland as runoff will eventually collect in a linear flow of downhill-moving water called a stream.

- The land area that contributes runoff to a stream is called its watershed.

- Water moves faster on the outside of a stream bend; this will cause erosion, which causes the stream to curve even more.

- Streams are straightened and their length decreased when very curvy sections are cut off.

- Stream erosion and deposition will both destroy and create land adjacent to the stream channel. This created land is very susceptible to flooding.

Overland Flow

Water that reaches the surface of Earth but does not infiltrate stays on the surface (by definition: if it does not go into the ground, then it must stay on top). Gravity will immediately act to bring the water closer to the center of Earth. So water that does not infiltrate will move downhill across the ground. Water moving like this is called runoff.

All overland flow of water eventually reaches a low point, with only one way down. Water always will follow the steepest path of least resistance as gravity pulls it downward, so most of the time moving water will follow only one path (typically the steepest). Consider a case where you are standing on the side of a hill. There are many potential paths downward from your position on the hillside. One may be steeper, or better than the others, but there are many routes you could take that would move you to a lower elevation. Now imagine that you are standing in the middle of a small stream. If you face the direction in which the water is moving, you will be looking downhill. If you face *any* other direction, the ground will run uphill away from you. There is only one way down in a stream, the direction in which the water is moving.

A line of moving water, being pulled downward across Earth, is

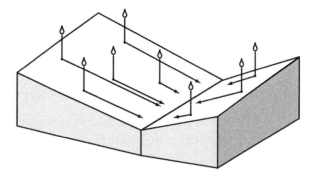

Figure 13.1. Water lands on the surface of Earth, and if it does not sink in, it immediately moves downhill, across the ground, under the influence of gravity.

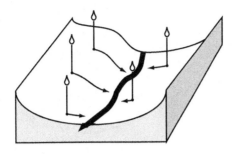

Figure 13.2. Once the runoff water reaches a low point, it joins in the line of water flowing downward from points higher up the line. At this point there is only one way in which the water can move.

called a stream. "Rivers," "creeks," "brooks," or "runs" are all names used to describe a stream—a linear flow of water. The various terms are used informally to denote the size of a stream. To a physical geographer, streams are streams. They are all linear flows of water moving downward under the influence of gravity; whether river or brook, they all follow the same rules.

It is possible for water to flow downhill into a body of water from which there is no outlet. Once the water enters a lake (or pond, or ocean, or even a puddle), it is no longer moving downward under the influence of gravity, and thus is not part of overland flow or a stream.

Once water runs off across the surface and ends up in a stream, how many possible routes are there for it to continue flowing downward?

Answer: Once water is in a stream, there is only one direction down.

Watersheds

Every stream and body of water has a surrounding area of ground called its watershed. A watershed is an area of ground that contributes runoff to a stream. A watershed sends runoff to only one stream.

Not only does the entire stream have a watershed, but also every point on the stream has a watershed that contributes runoff that flows past that point. Larger streams, such as the Mississippi River, are fed by many smaller streams, each of which has its own watershed. A lake or a

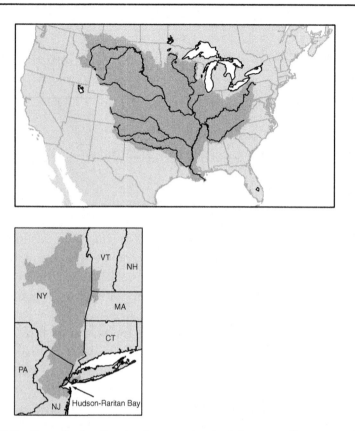

Figure 13.3. Every body of water has a watershed that contributes runoff to it. *Top:* Larger streams, such as the Mississippi River, are fed by many smaller streams, each of which has its own watershed. The watersheds of each tributary also are part of the watershed of the Mississippi. All the rain that runs off the land in the Mississippi's watershed will enter the Gulf of Mexico in Louisiana via the lower Mississippi River system. *Bottom:* The open-water bay around New York City receives water from the Raritan River, the Hudson River, and numerous smaller streams. All of the streams that feed into the bay have their own watershed. The cumulative watershed of all those small streams is the watershed for Hudson–Raritan Bay. Any drop of water that runs off from the bay's watershed will eventually end up in the bay. The path the runoff follows before it gets to the bay doesn't matter.

bay also can be said to have a watershed. Every drop of water that falls to Earth lands in the watershed of some water body. If that drop of water does not infiltrate, it will run off to that water body.

What is a watershed? _____

   Answer: It is the area of ground that contributes runoff to a body of water.

Water and Sediment Transport

Moving water can carry sedimentary particles with it. There are two critical concepts that will run through this chapter and carry over into other chapters.

1. The faster the water moves, the bigger the particles it can carry. If it goes slowly, it can carry only small pieces.

2. The faster the water moves, the more particles it can carry. If it goes slowly, it can carry only a small amount of material.

Faster flow = more pieces + bigger pieces

As water moves faster, what happens to its ability to mobilize and transport sediment? _____

Answer: Faster-moving water can carry more particles and bigger particles.

Stream Meandering

A stream rarely travels a perfectly straight path for long distances. Inevitably there are bends produced by the type of terrain: a tree, a rock, a fallen log. As the water in a stream arcs around an obstacle or a bend, it will move at slightly higher speeds on the outside of the turns and at a

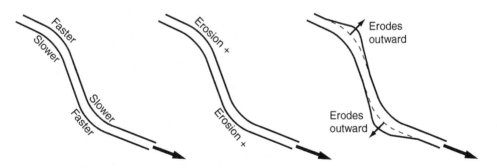

Figure 13.4.   *Left:* Water in a stream travels faster than average on the outside of a turn and it travels slower than average on the inside of a turn. *Center:* Water that moves faster has a greater ability to cause erosion than water that moves slower. *Right:* The greater ability of faster water to mobilize and transport sediment will cause the outside of a bend to be eroded.

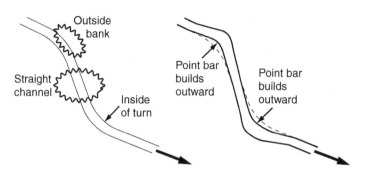

Figure 13.5. *Left:* Material eroded by fast-moving water upstream (on the outer banks of turns or in straight sections that precede turns) can be deposited *(right)* as point bars downstream if slower moving water on the inside of a turn doesn't have enough speed to keep transporting the sedimentary particles.

slightly slower rate on the inside. Because the water traveling on the outside of a bend is moving faster than the other water in the stream, over time the water on the outside of the bend will have more erosive capability. (The faster the water travels, the more material it can carry.)

This faster-moving water will erode the bank and deepen and extend the outside of the turn. Material on the outside of the bend is eroded and carried downstream. Some of that eroded/transported material will eventually end up being carried by the water to the inside

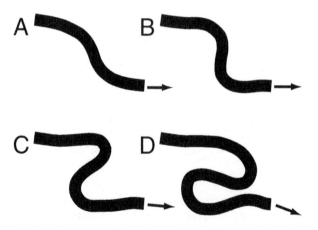

Figure 13.6. *A* and *B:* As time goes by, erosion pushing outward and deposition on the inside of turns will increase the sinuosity of a stream. *C* and *D:* As even more time goes by, sinuosity can greatly increase, and the stream will become much longer.

of a bend. Water on the inside of a bend moves slower relative to the rest of the stream. The slowed–down water is no longer able to transport as much material as it was carrying before it lost its speed. Therefore it must drop some of that material (deposit it). The inside of a bend will build up as material eroded by the fast-moving water upstream is deposited at the inner turns, where the stream is now moving slower.

Erosion on the outside of a turn and deposition downstream on the inside of another turn will cause a stream to become more sinuous over time.

Why does sediment accumulate on a point bar on the inside of a turn?

Answer: Water moves slower on the inside of a turn and is no longer capable of carrying the sediment that was being transported by faster-moving water upstream from the turn.

Cutoff Meander

As erosion keeps pushing the bends outward, and sections of a meander begin to get near one another, it may be possible for the stream to break

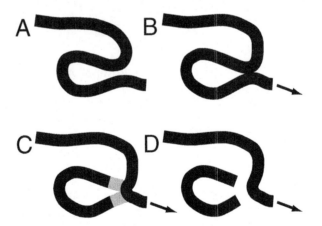

Figure 13.7. _A_ and _B:_ If erosion on the outside of bends keeps progressing, it is possible for the stream to break through from one section to another. _C:_ Water will follow the shorter, steeper path through the cut. Sediment being transported in the main channel will settle and deposit if the current carries it into the slow-moving section of the cutoff meander. _D:_ Eventually the parts of the old meander that are nearest the cutoff will fill with deposited sediment. The back section of the meander will be isolated and become an oxbow lake.

This montage of aerial photographs shows the Mississippi River near Greenville, Mississippi. A large oxbow lake (the dark area left center) is a remnant of a former river meander. Other abandoned courses of the river also are apparent, as are the agricultural and social uses of the floodplain. (Photograph base image courtesy of USGS via Terraserver.)

through a narrow neck separating its parts. If a breakthrough happens, it is called a cutoff. The main flow in the stream will follow the shorter, steeper path. It is a steeper path through the cutoff than around the old meander because it will be the same change in height but in a much shorter path. The stream will abandon its old channel for the new path, and the old part of the channel will become a slow-flowing backwater, "cut off" from the main flow of water.

Any water that still flows into the abandoned channel will be moving at a much slower rate than the water following the steeper route in the main channel. Eventually the parts of the abandoned channel closest to the cutoff will be filled in with sediment that the slower-moving water there is no longer capable of carrying. Most of the sediment that is carried into the cutoff meander will quickly settle in the area close to the main channel. The accumulation of deposited material will isolate the back portion of the abandoned channel and transform it into a lake. Because of its shape, the abandoned (now isolated) old channel is called an oxbow lake.

How is an oxbow lake separated from the main channel? _____

Answer: It is isolated when sediment is carried into the former channel and dropped when the water speed is too slow to continue transporting it.

Creation of a Floodplain

In the process of a stream meandering from a straighter path to a more sinuous route, an important thing happens: land is created and destroyed. As the outside of the stream cuts outward from its original course, it erodes the streambank. Eventually deposition on the inner bank will progress outward, and new dry land along the inside turn is built in the same place where old land was once eroded away.

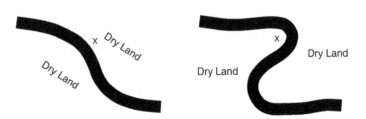

Figure 13.8. When a stream meanders from a straighter path *(left)* to a more sinuous path *(right)*, dry land is destroyed in some places and created in others. In this case, point X is destroyed by erosion and then restored by sediment deposition as dry land on the other side of the stream.

The inner bank builds out into the stream when sediment is deposited there by slower-moving water as it loses the ability to transport the sediment. The elevation of this new land will be limited by the height of the water in the stream. The stream can only drop sediment where the water is flowing.

During a flood, when water level exceeds its normal height, the water will move out of the channel onto the adjacent, newly created, low-lying land. Sediment transported by fast-moving floodwater in the

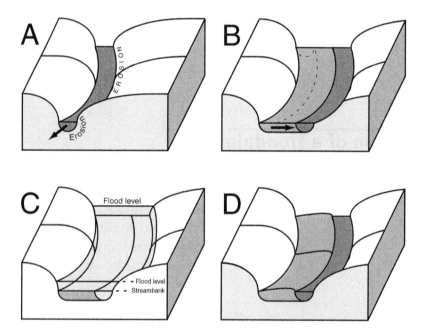

Figure 13.9. *A:* Water in a stream moves at a faster rate on the outside of the bend. This higher speed will promote erosion of the outer bank. *B:* As material is eroded from the outer bank, the inner bank will build outward because slower-moving water there has less ability to transport sediment. *C:* During a flood, a stream will overflow its banks, and the fast floodwater will carry sediment with it as it moves out of the channel. *D:* That sediment is deposited when floodwater spreads out and slows down. The elevation of the newly created land on the inside bend of the channel will increase as successive floods deposit layer after layer of sediment. However, the low-lying, dry land adjacent to the channel is constructed of sediment deposited by water in the stream and by floods, so it always will be below the level of the highest flood at that location. Therefore the floodplain always will be exposed to flooding.

Roberts Swamp Brook (flowing into the page, at left) in Brielle, New Jersey, is overflowing the bank onto its floodplain at right. (Photograph courtesy of MACGES.)

channel will be carried by the water onto the adjacent land and deposited when the water spreads out and slows down.

The stream overflows onto (floods) the flat, low-lying land (a plain) when there is too much water to be handled by the channel. Each successive flood carries sediment out of the channel onto the adjoining land, so the floodplain will get a little higher each time the stream floods. However, the overflowing water can never deposit sediment at an elevation higher than the floodwater. So the maximum elevation of the floodplain always will be below the highest flood levels. The floodplain always will be lower than the height of the maximum flood, so it always will be susceptible to flooding.

How does a floodplain get high enough to become dry land?

_____

Answer: Every time that water moves onto the floodplain, it can bring sediment and leave it behind to build up the ground elevation.

1. The area that contributes runoff to a body of water is called its
   _____.
   - a. watershed
   - b. floodplain
   - c. meander
   - d. cutoff

2. If a stream gets too sinuous, it will break through and create a
   _____, which will straighten the course of the stream.
   - a. watershed
   - b. floodplain
   - c. meander
   - d. cutoff

3. As water moves faster it can carry _____.
   - a. more sediment but smaller-size pieces
   - b. less sediment and smaller-size pieces
   - c. more sediment and larger-size pieces
   - d. less sediment but larger-size pieces

4. Water that lands on the surface of Earth and moves across it under the influence of gravity is said to be moving as _____.

5. A linear flow of water being pulled downward by gravity is a _____.

6. An oxbow lake forms after a stream is cut off because faster water in the main channel slows down and drops its sediment load when it moves into the old meander. (True or False)

7. Every body of water has a watershed. (True or False)

8. Why is a floodplain so susceptible to flooding?

9. How does the increase in the speed of water on the outside of a bend make a stream more sinuous?

10. Where does the material used to construct a point bar on the inside bend of a stream come from?

ANSWERS

1. a	2. d	3. c	4. runoff
5. stream	6. True	7. True	

8. It has been constructed of material carried by the stream, and its elevation is therefore below the maximum height that the water has ever reached in the past.

9. As water on the outside of a bend moves faster, it can more easily erode the streambank and extend the turn outward.

10. Material that is deposited in the point bar comes from sediment that was mobilized upstream. The slow-moving water on the inside of a bend is not competent to continue transporting sediment that faster waters carried downstream to that location.

Links to Other Chapters

- The amount of water moving in a stream and the frequency with which it floods are functions of a place's climate (chapter 7).
- Particles that are transported by a stream can be physically weathered by the process of rolling along or striking other objects (chapter 11).
- Because water in streams and ice in glaciers are each pulled downward by gravity (chapter 11), streams and glaciers often carve routes for each other (chapter 14).
- Precipitation can infiltrate into the ground before the precipitation has a chance to run off (chapter 12).
- Streams usually transition into an estuary mixing zone before they reach the open ocean (chapter 15).

<u>14</u> Wind and Ice

Objectives

In this chapter you will learn that:

- The wind can move sedimentary material in places such as deserts or saltwater beaches, where there is an absence of vegetation.

- The wind can move particles by suspending the smaller ones, rolling or sliding the larger ones, and bouncing the middle ones.

- If the wind is blocked or slowed, it can be forced to drop any particles it is carrying.

- Glaciers are formed in places where more snow falls in winter than melts away in summer.

- Glaciers move downward under the influence of gravity.

- Moving ice is a very powerful eroder, and the eroded material is pushed along by the glacier.

- A landform called a moraine marks the farthest advance of a glacier.

Wind and Ice

The two geomorphic agents discussed in this chapter, the wind and glacial ice, are restricted by environmental factors and can only be effective in particular places. The wind is an effective landscape force only in places where it can come into direct contact with bare ground. Practically, this restricts aeolian geomorphology (wind-constructed landscapes) to arid/desert areas and saltwater shorelines. Glaciers also are regionally restricted. Glaciers are only found where it is cold enough to sustain snow and ice year-round. Thus glaciers only affect the landscape in the high latitudes and high mountains. Climate (dry→wind, cold→ice) and terrain (beaches→wind, mountains→ice) are the two factors that limit these agents.

What role does climate play in the geomorphic ability of the wind?

Answer: The wind can only be effective in dry places where plant cover is limited and the moving air can come into direct contact with the bare ground.

The Wind

The wind is the only geomorphic agent that is directly produced by differences in atmospheric pressure. The flow of air is not produced by gravity, nor is it confined to a channel, but in one very important way moving air is like moving water. The wind can move material by pushing or rolling it, or by lifting it up off the surface and carrying it along. The same principles that apply to the velocity of moving water apply to the velocity of moving air:

• The faster the wind, the more material it can move.

• The faster the wind, the bigger the particles it can move.

While the wind operates nearly every place on Earth, it is only really important as a geomorphic agent in places where moving air can come into direct contact with the bare ground. If the ground is covered with plants, friction slows the wind speed, and stems and leaves break up air streams, so the wind cannot move sedimentary material. There are two kinds of environments in which the wind is an important geo-morphic agent.

Arid Regions

Deserts are areas where there is not enough water for plants to grow. If the quantity of available water cannot sustain vegetation, the wind can be effective.

Saltwater Shorelines

Salt water makes it difficult for plants to grow. Direct contact with the water or with salt spray from breaking waves can kill plants. A few coastal plants can handle salt, but they might not grow adjacent to the water, where strong currents flow or breaking waves release a lot of energy. Often there is an unvegetated zone along a saltwater shoreline where the wind can move sediment.

Why can't the wind effectively move sediment when plants are growing well? _____

> *Answer:*   The plants break up and interfere with the airstreams, so the wind cannot mobilize sedimentary particles.

Sediment Transport

The wind is not as strong as flowing water in transporting sediment. Moving water can mobilize and transport much bigger material than the wind can. For practical purposes the wind can only move pieces that are up to gravel size. Nearly all wind-transported sediment is sand size or smaller (smaller than 1–2 mm). The smallest particles can be lifted up into the air and carried along by the wind. These dust particles are tiny enough that turbulence in the air can keep them suspended against the force of gravity (which always acts to pull them toward the center of Earth). Of course, if the air becomes still, gravity will be able to pull dust out of suspension and back to the surface. The largest-size particles that the wind can move are too heavy to be lifted into the air. The wind will transport them by sliding or rolling them along the ground.

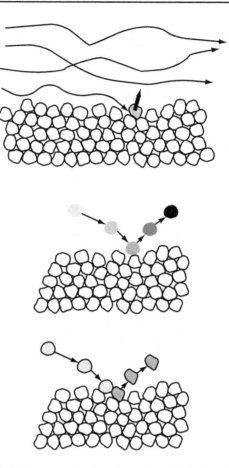

Figure 14.1. *Top:* Saltation can begin when the wind directly lifts a grain into the air. *Center:* A saltating particle can continue traveling along if it hits the ground and bounces back up into the air. *Bottom:* Saltation can be initiated when a particle that is being blown by the wind dislodges a different particle and forces it up into the air.

Mid-size particles can be rolled, too, or they can be transported by saltation. A saltating particle moves by a series of wind-aided jumps. The jump into the air can be initiated by:

- air turbulence that lifts the particles directly

- a bounce that reelevates an already moving particle

- an impulse from a moving particle that gets a different grain to move

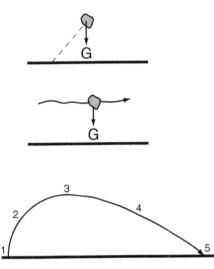

Figure 14.2. *Top:* Once a particle is up off the ground, gravity will immediately try to pull it down, toward the center of Earth. *Center:* As gravity acts to pull a saltating particle back to the ground, the wind will be carrying the particle along. *Bottom:* (1) The particle is initially propelled upward. (2) Gravity immediately acts to slow its rise—but the wind is beginning to move it along. (3) Gravity has stopped the particle's upward motion and is about to cause it to begin falling back to the ground, and the wind is still propelling it. (4) Gravity is pulling the particle back down, and the wind is still carrying it along. (5) The particle hits the ground (is out of the airstream), but its arrival back at the surface can trigger a bounce back up into the air, or it could impel another grain up into the wind.

Saltating grains are too large to be kept aloft by turbulence, so gravity will quickly be successful in forcing the particle back to the surface. However, in its brief time away from the surface the particle will be in the airstream. The air will carry it along until gravity pulls it back down, out of the wind. Thus if the particle can be lifted, it will follow an arcing, hopping path, and its arrival back at the surface can trigger a bounce back up into the air, or it could impel another grain up into the wind.

What are three ways that a sand grain can become airborne to begin a saltation hop? _____

Answer: A sand grain can be lifted directly off the surface by the wind, it can bounce off the ground after a prior saltation hop, or it can be propelled by impact from another saltating grain.

Dune Creation

Dunes are formed when sedimentary particles transported by the wind are deposited. Deposition occurs when wind speed is reduced to the point where the air can no longer transport sediment. If there is an area sheltered from the wind, any sand particle that falls into the quiet zone will remain there.

There are several types of dunes, but transverse dunes along ocean shorelines are one of the most common and most well-known forms. On a saltwater beach, plants will be at some distance from the water's edge. Plants need to be safe from storm waves and far from salt spray. With enough setback, some plants (usually grasses) can survive the harsh beach conditions—few nutrients, no shade, hardly any fresh water—and grow.

If the wind is strong enough to mobilize particles on the beach, then the particles can accumulate into a dune if they are transported to the grassy area. The grass can physically block the movement of particles, preventing them from moving on. The grass also can create a sheltered area with lower wind speeds, which can cause the wind to lose its ability to transport material. Sediment can build around the vegetation, and the accumulated material will create a dune.

As material accumulates in a dune, the new dune itself partially blocks the wind. In the area behind the dune that is sheltered from the wind, the air will no longer be able to continue transporting any sediment it has mobilized. If the particle finds itself in a protected, leeward area, it will be deposited. So any particles that are blown into a dune area can produce further dune growth.

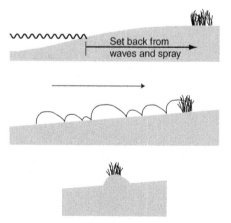

Figure 14.3. *Top:* On a saltwater beach, vegetation cannot grow too close to the shoreline because it needs to be set back from storm waves and saltwater spray. *Center:* Grass can be a direct obstacle that saltating grains can't leap past. Grass also will act to diminish wind speed, which also will promote deposition. *Bottom:* Grass can continue to grow up through the accumulated sand pile and enable more material to be deposited, which will enlarge the dune.

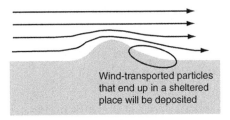

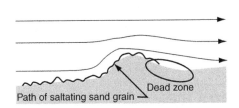

Wind-transported particles that end up in a sheltered place will be deposited

Figure 14.4. *Top:* If a dune disturbs the flow of the wind, or if transported particles move into a sheltered place, sediment can be deposited. *Bottom:* A saltating particle that is being carried over a dune by the wind can be dropped if the dune form disturbs the wind field.

Fences are installed along the Manasquan, New Jersey, beach to reduce the transport of windblown sand. Keeping sand on the beach instead of having it blow away inland provides a measure of storm protection. (Photograph courtesy of MACGES.)

Other objects, particularly fences, also can disturb the wind speed or wind path and are purposely used to cause windblown material to accumulate. Fences also can physically block the movement of sedimentary particles.

How does grass or a fence cause windblown sediment to accumulate into a dune? _____

Answer: Grass or a fence decreases the wind velocity and causes the wind to deposit any material it can no longer carry.

Accumulation of Snow and Ice

Ice is a denudational agent that can only be effective in regions where the temperature gets below freezing: mid- to high latitudes and high elevations. Ice can physically weather rock because of the force it exerts when water freezes and expands. Ice also can weather and transport rock and sediment in places where it stays cold enough to create glaciers.

Glaciers can be created in places where more snow falls in the cold part of the year than melts away in the warmer part of the year. If this annual accumulation of snow goes on for many, many years, eventually the snow on the ground will get quite thick. The weight of all the overlying material puts pressure on the lower snow and transforms it into ice. If a snowball is squeezed very hard, it can turn into an iceball. A snowy sidewalk can get icy if people walk on it. A plowed road needs sand or salt to keep it from becoming a sheet of ice. An ice form cre-

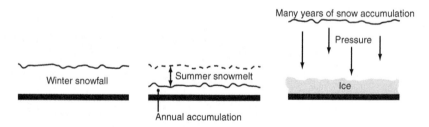

Figure 14.5.   *Left to right:* If more snow falls in the cold season than melts in the warm season, there will be a yearly accumulation of snow. The weight of snow that accumulates over many years will put pressure on the lower layers, which can transform the bottom snow into ice.

ated by the accumulation of pressurized snow leads to formation of a glacier.

What kind of climate condition is necessary for the formation of glaciers? _____

> *Answer:* More snow has to fall in the cold season than melts away in the warm season to allow snow to accumulate into a glacier.

Moving Ice

Glacial ice can move. It can distort and flow in response to outside forces. By far the most important outside force on a glacier is gravity. Just as it does with everything else, gravity will pull the glacier toward the center of Earth. If the ice can move downhill (closer to Earth's center), it will begin to flow in that direction—although it moves very, very slowly.

Thick ice, moving downward, is a very powerful weathering and erosion force. The front of the glacier acts like a bulldozer. It pushes along everything it encounters. Rocks and dirt picked up and carried along by the glacier work like sandpaper to scour and abrade any rock the ice moves over. The ice also can pluck chunks of rock out of the terrain it is moving across. All of this bulldozed, scraped, and plucked material is pushed downward by the ice. This pile of debris contains all sorts of things—basically everything the glacier has run into. Rocks will

The Margerie Glacier in Glacier Bay National Park flows downward until it meets open water in the Tarr Inlet. The black streaks in the ice are eroded rock and soil that the ice is transporting. (Photograph courtesy of P. Craghan.)

Figure 14.6. If gravity pulls the ice in a glacier downhill, the ice will slowly move along, scraping and bulldozing everything in its path.

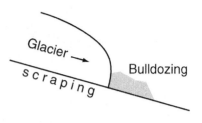

come in all sizes and mineralogies; there will be clay, silt, sand, gravel, and larger particles; plants and trees can be mixed in, too. The glacier erodes the landscape and carries that material to a lower elevation as gravity pulls the ice downward.

How do glaciers erode the terrain they move across? _____

Answer: They scrape the terrain they move across, bulldoze material in front of the moving ice, and pluck sections of rock out of their matrix.

Glacial Melting

As a glacier moves, eventually it will reach an area where temperatures are too warm to sustain the ice. A glacier will extend forward until it reaches a location where ice melts faster than it can be replaced by the moving glacier. This could be at a lower elevation on a mountain, or a warmer climate zone.

If global climate change raises the temperature, then a cold place that once sustained a glacier can be transformed into a not cold enough place. As warming temperatures affect the ice, melting will give the

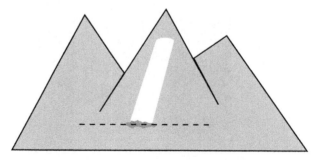

Figure 14.7. The downward extent of ice depends on how fast the glacial ice melts as it moves in a warmer climate zone.

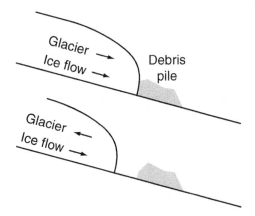

Figure 14.8. The glacier is the moving ice formation itself. Melting will cause the ice in a warm area to melt; thus the glacier will cover a smaller area. The glacier looks like it is moving backward, out of the too-warm area it has moved into, but it is just losing melted-away material from its front. The ice itself only moves because of gravity. Ice will not move uphill. In fact, the ice in a glacier will continue to move downward by gravity even as the glacier is losing volume due to melting.

appearance that the glacier is retreating backward. The area covered by glacial ice can shrink back toward a colder zone that can sustain ice. However, the ice in the glacier cannot flow backward. Ice will only move downward under the influence of gravity. Ice will not flow uphill.

Explain how a glacier can move uphill while ice can only flow downhill.

Answer: Ice can only move downhill because gravity causes its movement. However, the area that is covered with glacial ice can decrease if warming melts the parts of it that are in places too warm to sustain ice.

Glacial Landforms

When a glacier retreats, it leaves behind evidence of its former presence. U-shaped valleys and moraines are indications that glaciers were present. Because ice is an effective eroder along all surfaces, a glacier will carve a U-shaped valley as it moves from a higher elevation to a lower one.

The pile of debris that a glacier pushes along when it is moving forward will be left in place after the ice melts away. This bulldozed moraine ridge, made up of all sizes and materials, marks the farthest point to which glacial ice ever flowed before it melted back.

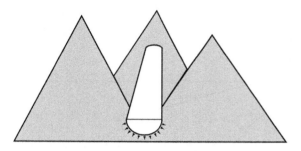

Figure 14.9. Ice will very efficiently erode rock along all of its contact surface. A U-shaped valley is evidence that there was once a glacier at that spot.

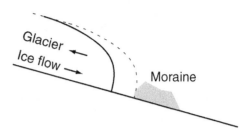

Figure 14.10. The bulldozed remains of everything the advancing glacier encountered will be left as a moraine landform after the glacier melts back from its farthest advance.

This U-shaped valley in Rocky Mountain National Park, Colorado, is a particular landscape characteristic indicative of past glaciation. (Photograph courtesy of National Geophysical Data Center.)

What is a moraine? _____

Answer: A moraine is a landform made up of all kinds of materials that were bulldozed by moving glacial ice. The moraine shows the farthest extent of the glacier's movement.

SELF-TEST

1. What kind of particles are most likely to be transported by aeolian saltation?
   a. dust
   b. sand
   c. gravel
   d. boulders

2. The wind is an important erosive force in forested areas. (True or False)

3. The wind can transport sediment to a higher elevation than where the particles started. (True or False)

4. A U-shaped valley is an indication that there once were glaciers in that spot. (True or False)

5. A mixed-up pile of sedimentary debris that marks the farthest extent of a glacier is a _____.

6. A _____ is a sedimentary feature that is created by the deposition of windblown material.

7. The two general types of places where the wind is an effective geomorphic agent are _____ and _____.

8. The two general types of places where glaciers can be found are _____ and _____.

9. How does airspeed affect the ability of the wind to transport material?

10. What causes snow to become glacial ice?

ANSWERS

1. b   2. False   3. True   4. True

5. moraine or terminal moraine or end moraine

6. dune

7. deserts; saltwater shorelines

8. high elevations; high latitudes

9. The faster the wind speed, the more particles the wind can carry, and the larger the particles the wind can move.

10. Snow can be transformed into glacial ice when so much snow accumulates that the overlying material pressurizes the low-lying snow into ice.

Links to Other Chapters

- A place with a dry desert climate is one location where the wind is effective (chapter 7).
- Glaciers can exist only where it is cold enough (at high latitudes and/ or high altitudes) for snow to last all year without melting completely (chapter 7).
- Mountains and high elevations are produced through tectonic processes (chapter 9).
- Sand being transported via saltation can "sandblast" and physically weather rock (chapter 11).
- Moving ice is a powerful weathering agent (chapter 11).
- Aeolian and glacial processes can both cause soil erosion (chapter 11).
- The paths that moving water and moving ice follow are closely related because gravity is the force that mobilizes each (chapters 11 and 13).
- A stream also transports sedimentary particles by suspension or saltation (chapter 13).
- The waves and tides that act on saltwater shorelines make those places likely to have aeolian transport (chapter 15).

15 Waves and Tides

Objectives

In this chapter you will learn that:

- Waves are energy moving from one place to another.

- Nearly all ocean waves are generated by the wind.

- When waves enter shallow water, they refract to approach the shore-line more directly.

- When the water is too shallow to support the waveform, the wave breaks and releases the energy that created it.

- If a wave breaks at an angle to the shoreline, the wave will generate a longshore current that moves parallel to the shore.

- People can interfere with the longshore current by building structures called groins.

- Tides are the cyclic raising and lowering of water levels.

- Tides are caused by the Moon, the Sun, and the rotation of Earth.

Waves

Waves are one of the most important processes in the coastal zone, especially on open shorelines. A wave is simply movement of energy. And it is the delivery of energy to the shoreline via waves that makes them so important. The word "wave" can be applied to a wide range of phenomena:

radar waves	microwaves	earthquake waves	light waves
radio waves	sound waves	shock waves	

All of these "waves" are energy moving from one place to another. The energy that moves as a wave across water is almost certain to have come from the wind. However, it is possible for other sources of energy to produce a wave. Boats make waves called wakes; earthquakes can generate tsunamis; throwing a rock into a pond will produce ripples. Nevertheless, the moving air in the atmosphere is responsible for almost all the transfer of energy into the water.

What is a wave? _____

Answer: A wave is movement of energy from one place to another.

Wave Creation

As the wind moves across the surface of the water, some of the wind's energy gets transferred into the water. The more energy the wind transfers to the water, the bigger the waves will be.

- The faster the wind blows (wind speed), the more energy the wind has, and the bigger the waves it can generate.

- The longer the length of time the wind blows (wind duration), the greater the amount of time it can transfer energy to the water, and the bigger the waves it can generate.

- The greater the distance over the water that the wind travels (fetch), the more opportunity there is for air–water interaction, and the bigger the waves that can be created.

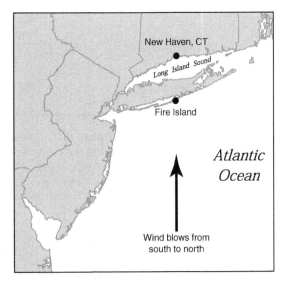

Figure 15.1. Fetch is the distance that wind blows over water. If the wind moved from south to north, then the fetch at Fire Island would be as much of the Atlantic Ocean as the wind crossed. For the same regional wind, at New Haven the fetch could only be the distance across Long Island Sound.

Thus, high winds blowing across a long length of water for a long time can transfer a lot of energy, which will move as very large waves. Once waves are generated, they move across the surface of the water until they encounter resistance.

What are three factors that influence how much energy moving air can transfer to water to generate waves? _____

Answer: wind speed, wind duration, and fetch

Refraction in Shallow Water

Waves first encounter resistance from the sea bottom as they move from where they were generated into shallow water. The wave length is the distance from the crest of one wave to the next crest. If the water depth is more than half of the wavelength, the wave is unimpeded by the seafloor. When the depth is approximately half of the wavelength distance,

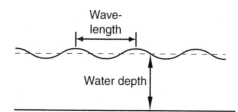

Figure 15.2. Wavelength is the distance from the crest of one wave to the crest of the next wave. When the water depth is less than half of the wave length, the wave is affected by the bottom.

the wave begins to be affected by the bottom. The shallower the water gets, the more that interference from the bottom will slow down the speed of the wave. So as water depth decreases, wave speed decreases.

If waves from deep water approach a shoreline at an oblique angle, different sections of the same wave will be in different depths of water at the same moment. The portion of the wave that is closest to the shoreline is in shallower water than parts of the wave that are farther off-shore. Waves move slower in shallower water than they do in deep water because they are being affected by the bottom. Therefore, different parts

Near Provincetown at the north end of Cape Cod, waves approaching from the Atlantic Ocean to the east *(right)* are refracted by shallow water so that they break at an angle that is much more head-on to the north-facing shoreline than their initial angle of approach. (Original photograph base image courtesy of USGS via Terraserver; the water north of Cape Cod was digitally enhanced to highlight the waves.)

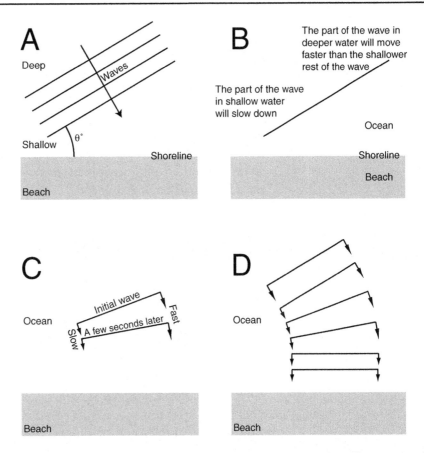

Figure 15.3. *A:* Waves move into shallower water at an angle as they approach land. *B:* The water depth will vary at different locations along the wave. The off-shore sections of the wave in deep water move faster than the parts of the wave that are closer to the shoreline in shallower water. *C:* Therefore the deeper section will travel a greater distance in a given amount of time. *D:* The difference in speeds will turn the wave until all parts of it are in the same depth of water (and thus all moving at the same speed).

of the wave will be moving at different speeds. As the wave continues to approach the shore, the slower-moving shallower end will not go as far in a given amount of time as the faster-moving deeper part of the wave will. The difference in speed between the deeper and shallower parts of the wave will cause the wave to turn toward the shoreline.

Once the wave turns to come straight into the shoreline, all its segments will be in the same water depth at the same time, and the wave will no longer turn. Turning due to the effects of water depth is called refraction. Refraction will cause waves to approach the shoreline more head-on than their initial angle of approach.

If a wave is approaching the shoreline at an angle, what makes it turn to strike the shoreline more directly? _____

> *Answer:*   The interaction of the wave with the bottom will slow down the wave speed. The part of the wave in shallower water will slow down, which allows the portion of the wave in deeper water to swing around and realign the wave's direction of movement.

Breaking Waves and the Longshore Current

If the wave moves into water that is too shallow to support its form, the wave will break. When a wave breaks, it releases back into the environment the energy that created it. Once a wave releases its energy, that energy can do work. The wave's released energy can physically weather rock or other shoreline material, or mobilize or transport coastal sediments.

Often an approaching wave will not have enough time to completely refract and hit the beach head-on. In that case the wave will be coming to the shoreline at an angle when it breaks and releases its energy. Because the wave is moving at an angle to the shoreline, the breaking wave will generate a current that flows parallel to the beach in a direction derived from the wave's approach. This water moves along the beach and is called the longshore current. The longshore current (like all moving water) can mobilize and carry sediment if the particles are small enough. The longshore current also is known as the littoral current and as littoral drift.

The longshore current's ability to mobilize and transport material can

Figure 15.4. When depth gets too shallow for the wave, it breaks and releases its energy.

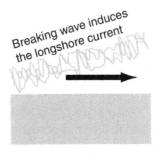

Figure 15.5. The longshore current is a shore-parallel movement of water set in motion by waves that break at an angle to the shoreline. The longshore current moves parallel to the beach and can carry sediment with it.

be strengthened by energy available from breaking waves. Energy released from breaking waves can lift material off of the seafloor and thus assist the current to mobilize and carry material that would otherwise be too heavy.

What happens to the energy that a wave represents when a wave breaks? _____

Answer: The energy is released into the environment, where it can do work.

Groins

Sometimes if a lot of sediment is being carried by the longshore current and the shoreline is experiencing a lot of erosion, people will build a structure to interfere with the drift. A groin is a shore-perpendicular structure designed to intercept sand moving along with the longshore current. Groins can be made of anything, but usually are made of wood or boulders. A groin has to be strong enough to withstand breaking waves and longshore currents.

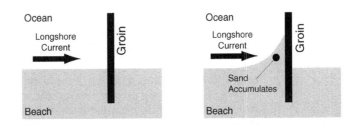

Figure 15.6. *Left:* A groin is a shore-perpendicular structure designed to stop beach erosion by intercepting sand moving in the longshore current. *Right:* A groin will block the longshore current and cause any moving sediment to accumulate.

Sand transported by a longshore current that cannot get past a groin will accumulate on the side of the groin that faces into the current. Groins can be very effective at interfering with the longshore current and trapping moving sand; this also is one of their problems. Groins prevent sand from moving on to other places, and that can cause erosion farther down the beach. Sand being transported by the longshore current can't get past the groin and accumulates on the upstream side. But the longshore current can mobilize and transport sand on each side of the groin. So on the downstream side of the groin, sand is eroded by the longshore current. Over time there is a net loss of material downstream from the groin because the sand carried away from the downstream side is not replaced by sand moving in from upstream.

A groin will cause sand to accumulate on its upstream side (and groins can effectively solve an erosion problem there). But because

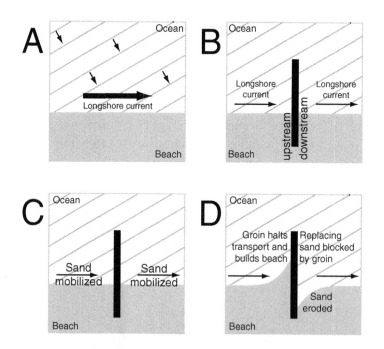

Figure 15.7. *A:* Waves approach a shoreline at an angle and create the longshore current. *B:* If a groin is present, waves still approach the shore on each side of the structure, so a longshore current is created on each side of the groin. *C:* The current can mobilize and transport sand on each side of the groin. *D:* A groin will interfere with the longshore current and cause deposition on the side that faces into the current. There will be erosion on the other side because material taken away by the longshore current is not replaced by new material.

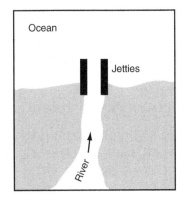

Figure 15.8. Jetties are intended to keep sand being transported by a longshore current from shoaling up the mouth of a river.

groins don't add sand to the system—they just affect its distribution—if sand accumulates in one place, there must be a compensating loss somewhere else.

A jetty is similar to a groin but has a different purpose. A groin is a shore-perpendicular structure designed to intercept sand moving

This photograph looks south along the New Jersey coastline. The Manasquan River inlet is protected by two jetties. Note how sand accumulated on the south side of the inlet and built a wide beach in Point Pleasant, New Jersey, while there is a considerable offset to the position of the beach in Manasquan, New Jersey, on the north side of the jetties. Several groins are visible along the beach in Manasquan. (Photograph courtesy of MACGES.)

in the littoral current to stop erosion and build a beach. A jetty is a shore-perpendicular structure designed to intercept sand moving in the littoral current to keep the sand from blocking an inlet and impeding shipping.

How does a groin cause a beach to accumulate sand? _____

Answer: A groin interferes with the longshore current, and any sand that is being transported by the current is blocked from further movement along a beach.

Tides

In addition to waves that bring energy from elsewhere to the shoreline, tides also provide energy at the coast. They do this by changing the level of the water in "high" and "low" cycles. Tides are caused by five things:

1.   The Moon's gravity affects large bodies of water and pulls water toward the Moon.

2.   Earth's monthly revolution with the Moon pushes water away from the Moon to the opposite side of Earth.

3.   The Sun's gravity affects large bodies of water and pulls water toward the Sun.

4.   Earth's yearly revolution with the Sun causes water to get pushed to the side of Earth that faces away from the Sun.

5.   The daily rotation of Earth on its axis spins any given place into line with the Moon twice a day.

These five things combine to raise and lower water levels in a predictable pattern called tides. The height difference between high water and low water is called the tidal range.

What is tidal range? _____

Answer: Tidal range is the height difference between high-tide water level and low-tide water level.

Top: At high tide, water inundates this cove and surrounds the pier. *Bottom:* At low tide, the water level has dropped considerably. The pier is now surrounded by a mudflat, exposed when the tide went out. (Photographs courtesy of MACGES.)

Moon Cycles and Spring and Neap Tides

The Moon's effects on large bodies of water are two to three times stronger than the Sun's, so the Moon controls when it will be high tide or low tide. High tide always will be aligned with the Moon, but the Sun will add or subtract from the total effect of the Moon to influence the tidal range. Most places have two high tides and two low tides per day. Sometimes local shoreline arrangements will disturb this pattern. Every point of Earth will spin around to be:

- on the Earth–Moon line twice a day = high tide

- perpendicular to the Earth–Moon line twice a day = low tide

The orbits of Earth, the Moon, and the Sun cause these three bodies to be lined up twice each month. This will cause tidal forces to be stronger and will generate what are called spring tides. The two times

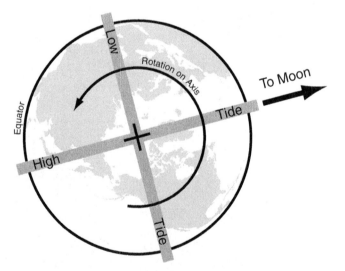

Figure 15.9. When a place, due to Earth's spin, ends up on the Earth–Moon line (as in the "high tide" box in this figure), it will be high tide because the Moon is pushing or pulling water to those areas. When a place spins around to be perpendicular to the Earth–Moon line (as in the "low tide" box in the figure), it will be low tide.

when Earth, the Moon, and the Sun are lined up are known as the full Moon and the new Moon.

At two other times each month, the Moon, Earth, and the Sun will be in perpendicular lines. When the Sun and the Moon are 90° apart with respect to Earth, it is called the first or the third quarter Moon. When the Sun and the Moon are working in different directions, tidal forces are weakened, and "neap tides" are the result.

New Moon

The new Moon occurs when the Moon is between Earth and the Sun. The side of the Moon that faces Earth is dark. (If there is absolutely perfect alignment in three dimensions, there will be an eclipse of the Sun at the exact instant of alignment.)

At the new Moon (and at all other times):

- The Moon's gravity pulls water toward it (Moon +).

- The Moon's orbit pushes water to the side opposite the Moon (Moon +).

- High tides will be in the two places where the Moon sends water.

- Low tides will be 90° off the Earth–Moon line.

- The Sun's gravity pulls water toward it (Sun +).

- The Earth–Sun orbit pushes water to the side opposite the Sun (Sun +).

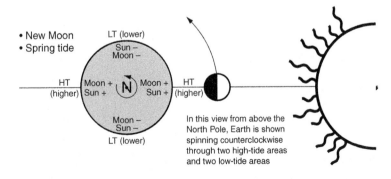

Figure 15.10. New Moon, spring tide.

- Earth rotates, so every day a place spins through two high tides and two low tides.

The Sun will be sending water to the same places (Sun +) that the Moon is (Moon +) and taking water from the same places (Sun −) that the Moon is (Moon −), so the Sun will reinforce the Moon's effects.

Because the Moon and the Sun are "working together," high tides will be a bit higher than average (Moon + and Sun +), and as a result, low tides will be a bit lower than average (Moon − and Sun −). These are *spring tides.*

First Quarter Moon

The first Quarter Moon occurs about one week after the new Moon, when the Moon has completed a fourth of its orbit around Earth. The Moon, Earth, and the Sun form a right angle (90°). From Earth, the Moon appears to be half sunlit and half in darkness at this time.

At the first Quarter Moon (and at all other times):

- The Moon's gravity pulls water toward it (Moon +).

- The Moon's orbit pushes water to the side opposite the Moon (Moon +).

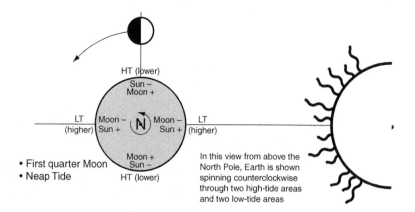

Figure 15.11.   First quarter Moon, neap tide.

- High tides will be in the two places where the Moon sends water.

- Low tides will be 90° off the Earth–Moon line.

- The Sun's gravity pulls water toward it (Sun +).

- The Earth–Sun orbit pushes water to the side opposite the Sun (Sun +).

- Earth rotates, so every day a place spins through two high tides and two low tides.

The Sun will be sending water (Sun +) to the places where the Moon is taking it from (Moon −) and taking water (Sun −) from the places where the Moon is sending it (Moon +). Thus, the Sun will work against the Moon's stronger tidal effects.

Because the Moon and the Sun are "working in different directions," high tides will be a bit lower than average (Moon + and Sun −), and as a result, low tides will be a bit higher than average (Moon − and Sun +). These are *neap tides*.

Full Moon

The full Moon occurs about two weeks after the new Moon, when the Moon has completed half of its orbit around Earth. Earth is in between the Moon and Sun. The side of the Moon facing Earth is completely sunlit. (If there is absolutely perfect alignment in three dimensions, then there will be an eclipse of the Moon at the exact instant of alignment.)

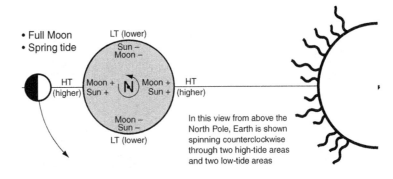

Figure 15.12. Full Moon, spring tide.

At the full Moon (and at all other times):

- The Moon's gravity pulls water toward it (Moon +).

- The Moon's orbit pushes water to the side opposite the Moon (Moon +)

- High tides will be in the two places where the Moon sends water.

- Low tides will be 90° off the Earth–Moon line.

- The Sun's gravity pulls water toward it (Sun +).

- The Earth–Sun orbit pushes water to the side opposite the Sun (Sun +).

- Earth rotates, so every day a place spins through two high tides and two low tides.

The Sun will be sending water to the same places (Sun +) that the Moon is (Moon +) and taking water from the same places (Sun −) that the Moon is (Moon −), so the Sun will reinforce the Moon's effects.

Because the Moon and the Sun are "working together," high tides will be a bit higher than average (Moon + and Sun +), and as a result, low tides will be a bit lower than average (Moon − and Sun −). Again, these are *spring tides.*

Last Quarter Moon

The last quarter Moon (a.k.a third quarter Moon) occurs about three weeks after the new Moon, when the Moon has completed three-

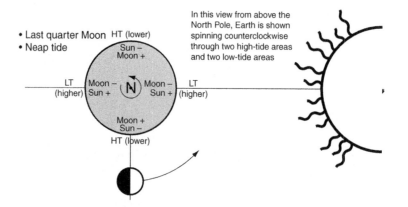

Figure 15.13.   Last quarter Moon, neap tide.

Moon Stage and Tidal State

Moon Stage	Moon–Earth–Sun		Condition	Compared to Monthly Averages		
				High Tide	Low Tide	Tidal Range
New Moon	In line	⬤●☼	Spring tide	Higher	Lower	Bigger
First Quarter	90°	●—☼	Neap tide	Lower	Higher	Smaller
Full Moon	In line	●⬤—☼	Spring tide	Higher	Lower	Bigger
Third Quarter	90°	—⬤—☼	Neap tide	Lower	Higher	Smaller
New Moon	In line	⬤●—☼	Spring tide	Higher	Lower	Bigger
First Quarter	90°	—⬤—☼	Neap tide	Lower	Higher	Smaller
Full Moon	In line	●⬤—☼	Spring tide	Higher	Lower	Bigger
Third Quarter	90°	—⬤—☼	Neap tide	Lower	Higher	Smaller
New Moon	In line	⬤●—☼	Spring tide	Higher	Lower	Bigger

●, Moon; ⬤, Earth; ☼, Sun.

fourths of its orbit around Earth. The Moon, Earth, and the Sun form a right angle (90°). From Earth, the Moon appears to be half sunlit and half in darkness at this time.

At the last quarter Moon (and at all other times):

- The Moon's gravity pulls water toward it (Moon +).

- The Moon's orbit pushes water to the side opposite the Moon (Moon +).

- High tides will be in the two places where the Moon sends water.

- Low tides will be 90° off the Earth–Moon line.

- The Sun's gravity pulls water toward it (Sun +).

- The Earth–Sun orbit pushes water to the side opposite the Sun (Sun +).

- Earth rotates, so every day a place spins through two high tides and two low tides.

The Sun will be sending water (Sun +) to the places where the Moon is taking it (Moon −) and taking water (Sun −) from the places where the Moon is sending it (Moon +). Thus, the Sun will work against the Moon's stronger tidal effects.

Because the Moon and the Sun are "working in different directions," high tides will be a bit lower than average (Moon + and Sun −), and as a result, low tides will be a bit higher than average (Moon − and Sun +). Again, these are *neap tides*.

The orbits of the Moon and Earth set up a monthly pattern of spring and neap tides that change the heights of high and low tides. The highest high tides of a month will be at the full Moon and the new Moon.

What is a spring tide? _____

Answer: It is the time at the full and new Moons when the tide-producing forces of the Moon and the Sun are working in the same direction and when high tides are higher than normal and low tides are lower than normal.

Tidal Impacts

Tides matter for three geomorphic reasons (and they also strongly affect how people use coastal areas).

1. By changing water height, tides affect the elevation where other processes such as breaking waves, littoral currents, or wetting and drying of the land occur.

2. A flood current is created as water flows from the sea toward the land in the period from low tide to high tide. An ebb current flows from the coast out to the open sea in the period from high to low tide. These tidal currents (like all moving water) can mobilize and transport sedimentary material.

Salt-tolerant grasses grow in a coastal salt marsh. At low tide, a mudflat is exposed at the border of the marsh. As the tide rises, the mudflat and much of the vegetation will be flooded and under water. (Photograph courtesy of MACGES.)

3. Salt water is brought into freshwater areas by the flooding current. This helps create a biologically important area called an estuary, where fresh and salt water can mix.

What is an estuary? _____

Answer: An estuary is a tidal environment where freshwater and salty ocean water mix.

1. Tidal range is _____.
   a. the height difference between high tide at spring tide and high tide at neap tide
   b. the height difference between a high-tide water level and a low-tide water level
   c. the portion of the ocean that experiences tides
   d. oceans plus estuaries

2. Which factor has the most effect on water level changes at a place from hour to hour?
   a. Earth's rotation on its axis
   b. Earth's revolution around the Sun
   c. The alignment of the Sun and the Moon
   d. The orbit of Earth and the Moon

3. When a wave _____ it releases the energy that was used to create it.

4. A _____ is a shore-perpendicular structure intended to stop erosion or to build a beach.

5. The distance over a body of water that the wind blows is _____.

6. The Moon has a stronger effect on tides than the Sun. (True or False)

7. At spring tide, high tides will be higher than average and low tides will be lower than average. (True or False)

8. What are two ways by which shallow water affects the movement of a wave?

9. How can a breaking wave contribute to sediment transport?

10. What creates the flood current in a tidal system?

ANSWERS

1. b   2. a   3. breaks   4. groin

5. fetch   6. True   7. True

8. As a wave moves into shallow water, its progress is hindered by interaction with the bottom, so it slows down. If the wave moves into water that is too shallow to support its form, the wave will break.

9. As the wave break, it releases energy that can be used to mobilize sediment. Breaking waves also can set up a littoral current that can transport material parallel to the shoreline.

10. As the water level in the ocean rises in the time from low tide to high tide, water will flow from the ocean toward the shoreline or into an adjacent estuary.

Links to Other Chapters

- Seasonal changes in Earth–Sun alignments (chapter 1) have some minor tidal effects.
- Wind is the principal agent in wave creation (chapter 4).
- Extratropical cyclones and tropical cyclones can be very strong storms that generate powerful ocean waves (chapter 6).
- Continental margins where subduction is taking place tend to have narrow continental shelves, which can influence how waves and tides affect a shoreline (chapter 9).
- The release of wave energy at a shoreline can physically weather rock (chapter 11).
- Estuaries are transition areas where streams mix with oceans (chapter 13).
- The release of energy from strong storm waves and the run-up of water on the beach are factors in how closely vegetation can grow to the shoreline (chapter 14).

Appendix 1:
The Ancient Explanation
of Earth–Sun
Relationships

An understanding of the old Earth-centered view of the universe is helpful because it can be used to gain an appreciation of how the Sun and Earth change their alignments due to Earth's tilt and revolution. This obsolete explanation also has produced some terminology that is still in use.

Before it was accepted that Earth moved around the Sun, it was assumed that the Sun went around Earth. Even today, to observers on Earth it appears that the Sun moves around the planet every day. On a yearly basis, people don't perceive Earth's tilt; instead, they see the movement of the Sun.

This "Earth-centered" view of the universe has worked for thousands of years, but it needs some creative physics to really function. All of the Earth–Sun relationships in effect when the Sun is directly overhead at different latitudes still need to be satisfied. This requires the Sun to revolve around Earth in a slightly different orbit every day. In this outdated model, Earth stays stationary and does not spin. The daily rise of the Sun in the east and its set in the west are accomplished by the direction of the Sun's daily trip around Earth.

When the Sun is directly overhead at the equator at the March and September equinoxes, it would be orbiting Earth in the same plane as the equator. From the March equinox to the September equinox, when the point directly underneath the Sun is between the equator and the tropic of Cancer (23½°N), the Sun is said to be "in the Northern Hemisphere"—that is, it is closer to the North Pole than it is to the

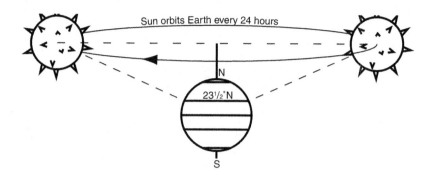

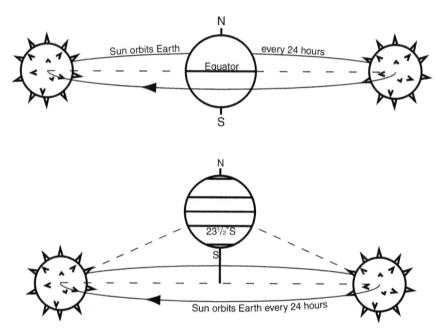

Figure A1.1. *Top:* At the June solstice, the Sun is said to be at its farthest north. It is directly overhead at the tropic of Cancer, 23½°N. *Center:* The Sun can be conceptualized as orbiting Earth from far above the equator at the March and September equinoxes. The Sun has to orbit in this direction for it to rise in the east and set in the west. *Bottom:* The Sun is at its farthest south, at the December solstice. It is directly overhead at the tropic of Capricorn, 23½°S.

South Pole. In the time from the September equinox to the March equinox, when the point directly underneath the Sun is between the equator and the tropic of Capricorn (23½°S), the Sun is said to be "in the Southern Hemisphere"—that is, it is closer to the South Pole than it is to the North Pole.

Appendix 2: Coriolis Force

Rotating Systems

The Appendix will try to qualitatively and mathematically explain where Coriolis force comes from and how it influences movement on the rotating Earth. Coriolis force comes into effect when an item moves across another rotating object. In geography, the item typically is moving across the surface of Earth, which rotates once around its axis every day. Earth is so large that we do not perceive the rotation. We can see the effects of rotation when the Sun and the Moon rise and set, or the stars move across the sky. The terminology itself gives it away: we assign the movement to the objects, not to Earth's rotation. Nevertheless, Earth does rotate on its axis, and this needs to be accounted for when things move across the planet.

Sometimes Coriolis is said to be an imaginary, or fictitious force. Because the Coriolis effect is not seen when observers are removed from the rotation (as they would be if standing on the Moon), the force does not exist in a fixed reference system. But within a rotating system, the Coriolis effect is very real. When you pick your coordinate system, you pick your forces. If geographic phenomena are observed from Earth, there is a rotating system and the Coriolis effect is in play. Mathematically, no coordinate system is "more correct" or "better" than any other; some are just easier to work with.

A Simple Demonstration

Earth is a difficult place to try to understand Coriolis.

- Earth is a three-dimensional, sphere-shaped object.

- Earth has gravity that keeps objects on its surface.

- Moving objects follow great circles unless other forces are applied.

- The Coriolis force is applied in different directions in the Northern and Southern Hemispheres.

It is much easier to understand why Coriolis force alters the paths of moving things if the system is simplified to a rotating disk, like a merry-go-round or a carousel.

Consider a large, two-dimensional disk that spins counterclockwise around a central axis of rotation. Andrew and Betty are directly opposite each other on the disk, near the outer edge. Chris stands off the disk, to its side. For our purposes, the disk is so large that Andrew and Betty do not perceive that it is spinning. As the disk rotates, at some point Andrew, Betty, and Chris will be in one line. Imagine at that moment that Chris tosses a ball over Betty's head toward Andrew. The ball sails over the disk as the disk continues to rotate. Each of the three people will perceive the flight of the ball a bit differently.

Chris, who is in a nonrotating universe, will see the ball fly straight

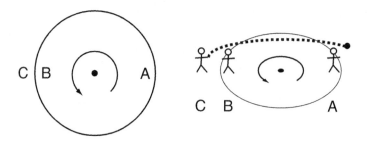

Figure A2.1. *Left:* This disk rotates counterclockwise around a central point. Andrew and Betty are across from each other on the spinning disk. Chris stands off to its side. *Right:* At the moment when the disk's rotation lines all three people up, Chris throws a ball over Betty, toward Andrew.

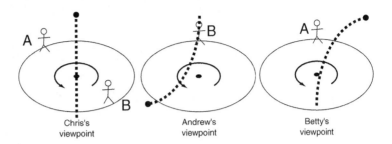

Chris's viewpoint

Andrew's viewpoint

Betty's viewpoint

Figure A2.2. *Left:* Chris, who is not on the disk, sees the ball fly over the disk as Andrew and Betty continue to spin around. *Center, right:* Andrew and Betty, who spin with the disk, see the ball fly over the disk, and they see the ball turn to its right, away from its initial path.

over the disk. As the disk rotates, Andrew and Betty spin away from the ball's flight path.

Andrew is on the rotating disk but is unaware that it is rotating. Andrew sees the ball after it flies over Betty's head into the disk. The ball is initially headed for him, but for some reason it turns to its right. Andrew blames the Coriolis force for making the ball turn from its path.

On the other side of the disk, Betty sees the ball after it passes over her head. It appears to be headed straight for Andrew, but then the ball makes a right turn and misses him. She blames the Coriolis force for making the ball turn to the right—changing its path.

Observers in a rotating system can perceive the same thing completely differently from observers who are not rotating.

Centrifugal Force

Newton's first law of motion says that an object in motion will stay in motion in a straight line at constant speed unless another force acts on it. (On Earth, friction and gravity are forces that can change the speed or direction of a moving object.) When an object is rotating on a disk, the object can be held in place by friction or other forces and thus will spin around instead of continuing in Newton's straight line. The object is "stupid"—it doesn't know what is happening to it; it just does what the forces make it do.

Consider the same large rotating disk from the previous section, but with only Andrew on it. The disk is so large that Andrew does not perceive that he is rotating. If Andrew is standing on a rotating disk, and he is holding a ball, the ball will go wherever Andrew carries it.

Let's consider what would happen if Andrew tossed the ball straight up as he was spinning around. The stupid ball does what the forces tell it to do, and the moment Andrew tosses the ball, the force that makes the ball spin with the disk is taken away. The ball will do what Newton said it would: the ball moves in a straight line in the direction it was headed at the moment it was tossed now that the force compelling rotation is removed. (The ball still will go up because it was thrown in that direction, and eventually it will momentarily stop rising and then be pulled down by gravity.)

An observer who is not on the disk would see the ball follow an arcing path. When the ball is released at position 0, it begins to move upward. At the same time, it will have an initial horizontal velocity that was imparted by the disk's rotation. So the ball will not merely go up, it also will move across in a straight line toward point X and will land at point X when its vertical motion brings it to the ground. In the time that elapses when Andrew tosses the ball upward at point 0 until the ball lands back down at point X, Andrew will have spun around to point 1.

When Andrew (who is rotating) tosses the ball, he will see it go up

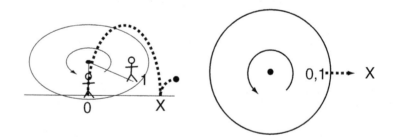

Figure A2.3. *Left:* After a ball is tossed vertically at point 0 it continues traveling horizontally in the direction it was going when it was released, until it falls back to the ground at point X. During the flight of the ball, Andrew, who tossed it, rotates from point 0 to point 1. *Right:* Andrew doesn't perceive that he is rotating, so for him point 0 and point 1 are identical. When he tosses the ball upward, he sees that it also flies out, away from the center of the disk.

and down, just as a nonrotating observer would, but instead of seeing the ball move forward, he will see it fly out away from the center of the disk. The force that pushes out is called a centrifugal force by the observer in the rotating system. This is similar to the force that makes you lean out from the center of a turn when an automobile veers sharply. This is a real force in the rotating system—it is not imaginary. You don't "imagine" that you (or your coffee) are being pressed toward the window when your car turns quickly. In a rotating system, centrifugal force pushes away from the axis of rotation. In a nonrotating system, centrifugal force is not needed to explain what is happening. A stationary observer would see your body (or your drink) continue moving straight even as your car turns off that path.

When an object in a rotating system begins to move outside the rotating context, it will have an initial velocity (inherited from the rotation) that will influence the path in which it travels. The ball does not go up and come back down to the same spot on the disk from where it was tossed. The ball doesn't remember that it was revolving around the disk. Depending on the coordinate system used, the ball either moves straight ahead (nonrotating reference) or straight away from the axis (rotating reference).

Coriolis Plus Centrifugal Forces

Consider a change to the earlier example, only now Betty tosses the ball across the rotating disk toward Andrew. As in the prior examples, the disk is so large that Andrew and Betty are oblivious to the fact that they are rotating. This case is different from the earlier one because instead of having a ball tossed by Chris, who is outside of the reference system, this time Betty, spinning with the disk, will toss the ball directly across the middle of the rotating disk.

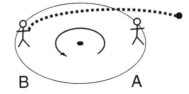

Figure A2.4. Betty intends to throw a ball directly across the rotating disk, toward Andrew. Each party is oblivious to the rotation of the very large disk.

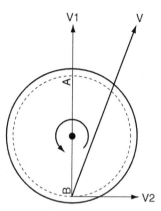

Figure A2.5. An observer who is not rotating will see Betty toss the ball across the rotating disk, toward Andrew. The velocity with which she throws the ball (V1) is added to the velocity that the ball inherited from the spinning disk (V2) to determine the direction in which and the speed at which the ball will travel (V).

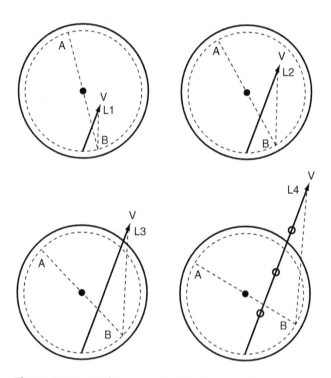

Figure A2.6. At four moments in time an observer not on the disk would see that the disk has rotated counterclockwise, taking Andrew and Betty with it. The ball moves along the path V to L1, L2, L3, and then L4. (The lines connecting Andrew with Betty, and Betty to the ball's position, are provided for reference to the following figure.)

At the moment Betty tosses the ball to Andrew, she removes the rotating influence from the ball. To a stationary observer, the ball will now travel in a straight line determined by the two influences acting on the ball at the moment Betty releases it. First, the ball was thrown by Betty toward Andrew. The ball will head in a straight line in the direction it was thrown with velocity V1. Second, the ball will have a horizontal velocity acquired from its rotational speed. The ball will continue to travel in a straight line with velocity V2, in the direction it was moving at the moment Betty released it. The two movements (V1 and V2) will combine to produce velocity V, which will determine the speed and direction of the tossed ball.

An off-disk observer will see the ball move along the V path as

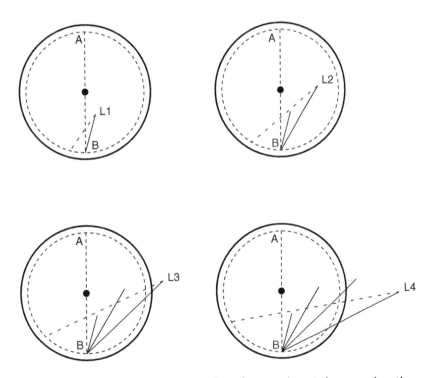

Figure A2.7. From Betty's perspective, she remains at the same location and Andrew always is directly across from her. The ball moves to position L1, which is to the right of the direction in which she threw it, and then it keeps turning to the right as it flies to L2, then L3, and then L4. Compare the lines connecting Andrew to Betty to Ln in this figure to the same lines in the previous figure.

Figure A2.8. Betty sees the ball continuously turning to the right after she throws it to Andrew. At one moment between L3 and L4, the ball is even moving straight away from Andrew, even though Betty threw it right at him. In the rotating system, there needs to be a force acting on the ball to make it turn. Betty assigns the turning influence to a Coriolis force.

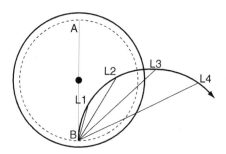

Andrew and Betty continue to rotate with the disk. At four equal moments in time the ball will fly to locations L1, then L2, then L3, then L4 while Andrew and Betty keep spinning.

In the rotating coordinate system, the same case is viewed differently. Betty, who is on the disk, sees that Andrew is always directly across from her, but she doesn't perceive that they are spinning. As she prepares to toss the ball, she faces Andrew and then throws it. However, instead of flying straight to Andrew, the ball mysteriously begins to turn to the right of the direction in which she threw it. As the ball keeps moving, it keeps turning even farther to the right of its flight path. At four equal moments in time after she lets go, Betty will see the ball at positions L1, L2, L3, then L4. Andrew, who never gets close to the ball, remains fixed in position across the disk from her.

From Betty's perspective, at four time steps she observes the ball at L1, L2, L3, and then L4. If we connect these points, we will note that Betty sees the ball continuously making a right turn. The ball appears to be flying to the right in an ever-expanding circular path. The ball is clearly not traveling in the straight path that Newton implied it would. However, Newton said an object would travel in a straight line if no other force acts on it. Betty knows Newton is right, so there must be another force. She says that the Coriolis force acts on a body in motion to make it turn. (She might say there is a centrifugal force, too, except that she doesn't know she is rotating.) Observers who are outside of the rotating system (like us) see the ball travel in a straight line from its initial position through point L4. They don't need to invoke Coriolis force to explain the location of the ball. In the nonrotating world, there is no Coriolis force acting on the ball. However, we have previously noted that observers in a rotating system will see things differently.

Gravity and Spheres

A rotating object would have to be very big for its occupants not to perceive the spin. Anything that large would have a fair amount of gravitational force. (For the moment, assume that Earth does not rotate; we'll add that component back in after we deal with gravity and great circles.) Gravity always is acting to pull things toward the center of Earth; therefore things moving across Earth won't be able to fly off the system, as they could with a two-dimensional disk. There is no edge to a sphere, and there is no top either. Thus an object cannot fall off of it. Movement will just make the object circle from one place on the sphere to another, and perhaps back to where it originated.

Figure A2.9.   Unlike a disk, a sphere does not have an edge. If gravity is strong, a moving object cannot fall off of a sphere.

Great Circles

A great circle is basically a circumference of a sphere. One such great circle is the equator, which divides Earth in half, and its plane passes through the center point of the planet. A straight line on a sphere may not follow the course you would expect it to. What makes sense on a map may not correlate to the real world. Maps of Earth are two-

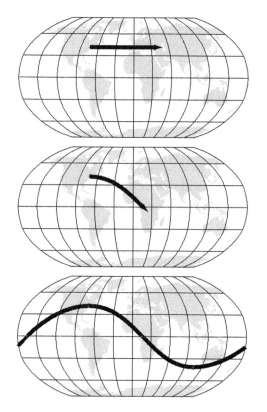

Figure A2.10. *Top:* The path that an object would follow on a flat map of the world if it were propelled due east from New York City. Remember, maps are simplifications—they are not real. *Center:* This is the path that an object propelled due east from New York would actually follow. Unless some other force acts on a moving object, it will travel in a straight line (which is a great circle on a sphere). *Bottom:* A full worldwide route that begins with a push due east from New York City would follow this path. Because a great circle is a circumference, the object will go around the world and return to its starting point in New York.

dimensional representations of this three-dimensional planet. On a flat map of Earth a great circle will be shown as a curving path. Remember that a great circle is a circumference that goes around the planet—it is actually a straight line on a sphere.

Think of an object that is set in motion on a nonrotating Earth. Assume that there is no friction to slow it down—once it begins moving, it can keep on going. After the object is given its initial push, if no other forces act on it, it will keep going in a straight line. Remember

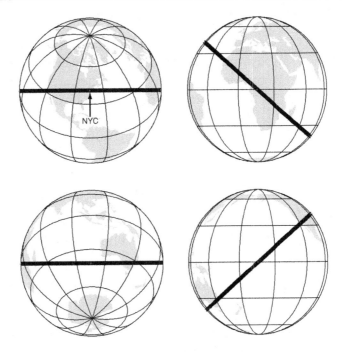

Figure A2.11. These four images of Earth show the great circle route that passes perfectly west to east through New York City. You can verify this yourself with some string and a globe. In these four images you can see how the great circle route is a circumference that divides Earth into two equal parts. Compare the straight line on a sphere with the curving path in the flat map of Figure A2.10. If you look at where the lines cross the continents you will see that the same route is mapped in each set of figures.

that a straight line on a sphere is a great circle (a circumference). In addition to the equator, all longitude lines (the meridians) are great circle paths.

Velocity on a Rotating Body

This section will show how velocity as observed from the rotating system can be correlated to velocity as observed from a fixed coordinate system. Some mathematics is employed, but it should be commonsense math.

Imagine a solid, sphere-shaped planet similar to Earth. It has the same dimensions, and it rotates on its axis with the same angular velocity. Imagine also that its inhabitants have painted the latitude and longitude lines on its surface so they are visible.

The formula to relate velocity in the two coordinate systems is:

$$V_{fixed} = dR/dt = dr/dt + (\omega \times R)$$

r = the vector from the origin of the *rotating* coordinate system to the point in question. If the origin is taken as the center of Earth, then the geometry will be greatly simplified. So dr/dt is the change in that vector over time, which is velocity in the rotating system.

ω = angular velocity; for Earth it is 2π radians/24 hours; the vector direction for ω points from the South Pole through Earth's center to the North Pole.

R = the vector from the origin of the *fixed* coordinate system to the point in question. If the origin of the fixed coordinate system is the center of Earth, then R is a radius from the center of Earth to the point on the surface in question, with a value of approximately 4,000 miles.

"$\times$" is a vector cross product. Because R, r, and ω are vectors, this is different from the symbol ordinarily used to indicate multiplication.

$(\omega \times R)$ = at the equator $(2\pi)(4000$ miles$)/24$ hours = 1047 miles per hour. The cross product direction is determined by the change in direction from the ω vector to the R vector. It will point from west to east and is tangent to latitude.

The velocity formula symbolically states that the velocity seen in the fixed system (V_{fixed}) is equal to the velocity seen in the rotating system (dr/dt) added to the velocity at which the object is rotating around on its axis ($\omega \times R$).

We are going to go through some cases to show how the velocity formula works. For the following cases, the observer in the rotating system would be someone on the planet, rotating with it. The observer in the fixed system would be someone off the planet (e.g., an astronaut in space) who can observe what is happening while being removed from the effects of the planet's spin. In each of the four cases an automobile

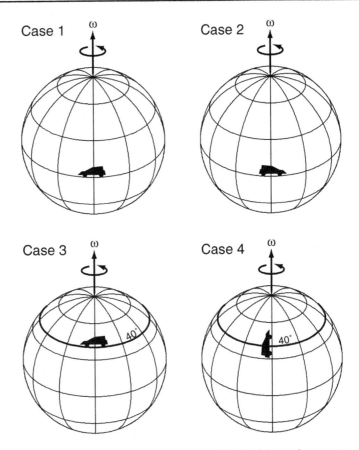

Figure A2.12. Case 1: An automobile is driven due west along the equator as the planet rotates from west to east. Case 2: An automobile is driven due east along the equator as the planet rotates from west to east. Case 3: An automobile is driven due west along latitude 40°N as the planet rotates from west to east. Case 4: An automobile is driven due north from 40°N for a short distance.

will be driven across the planet. If the automobile is said to go due east or west or north or south, that means it is driving along a visible line of longitude or latitude and the driver will steer the vehicle as necessary to stay on the geographic grid.

Case 1: A Car Is Driven West along the Equator

An automobile is moving very fast to the west along the equator when the driver notices that his speedometer reads 1,047 miles per hour.

$$V_{\text{fixed}} = dr/dt + (\omega \times R)$$

dr/dt = the velocity in the rotating system. In this case it is the speedometer reading, and it is negative because movement is from east to west.

$\omega \times R$ = 1047 miles per hour multiplied by the sine of the angle between ω and R, which at the equator is 90°

$$V_{\text{fixed}} = -1047 \text{ miles per hour} + (2\pi)(\sin 90°)(4000 \text{ miles})/24 \text{ hours}$$

$$V_{\text{fixed}} = 0$$

The rotating observer sees the automobile speeding at 1,047 miles per hour to the west. The fixed observer sees the vehicle remain in precisely the same spot. However, the fixed observer will see the planet rotate beneath the unmoving automobile.

Case 2: A Car Is Driven East along the Equator

An automobile is moving very fast to the east along the equator when the driver notices that his speedometer reads 1,047 miles per hour.

$$V_{\text{fixed}} = dr/dt + (\omega \times R)$$

dr/dt = the speedometer reading, which is positive because movement is west to east

$\omega \times R$ = 1047 miles per hour multiplied by the sine of the angle between ω and R, which at the equator is 90°

$$V_{\text{fixed}} = +1047 \text{ miles per hour} + (2\pi)(\sin 90°)(4000 \text{ miles})/24 \text{ hours}$$

$$V_{\text{fixed}} = 2094 \text{ miles per hour}$$

The rotating observer sees the automobile speeding at 1,047 miles per hour to the east. At this rate, the driver will circle the planet and return to the starting line in 24 hours. In contrast, the fixed observer sees the vehicle moving at 2,094 miles per hour around the sphere. To the fixed observer the vehicle will circle in 12 hours. The fixed observer also would note that the planet will have gone through only half of its rotation in that time.

Case 3: A Car Is Driven Due West at 40°N Latitude

An automobile is moving very fast to the west along latitude 40°N when the driver notices that his speedometer reads 802 miles per hour.

$$V_{fixed} = dr/dt + (\omega \times R)$$

dr/dt = the speedometer reading, which is negative because movement is east to west

$\omega \times R$ = the angular velocity multiplied by the sine of the angle between ω and R, which at 40°N latitude is 50° (sin 50° = 0.7660)

$$V_{fixed} = -802 \text{ miles per hour} + (2\pi)(0.7660)(4000 \text{ miles})/24 \text{ hours}$$

$$V_{fixed} = 0$$

The rotating observer sees the automobile moving at 802 miles per hour to the west (i.e., the value on the speedometer). The fixed observer sees the vehicle remain in precisely the same spot, and the fixed observer will see the planet rotate beneath the unmoving automobile.

Case 4: A Car Is Driven Due North at 40°N

At 40°N latitude an automobile is driven due north along a line of longitude at 60 miles per hour for 1 minute. For this case we will break the travel into two components (north/south and east/west) and then combine the two results to get our answer. First east/west:

$$Ve_{fixed} = dr_e/dt + (\omega \times R)$$

$dr_e/dt = 0$; there is no east/west movement in the rotating system, since the car is being driven northward

$\omega \times R$ = the angular velocity multiplied by the sine of the angle between ω and R, which at 40°N is 50° (sin 50° = 0.7660)

$$Ve_{fixed} = 0 \text{ mile per hour} + (2\pi)(0.7660)(4000 \text{ miles})/24 \text{ hours}$$

$$Ve_{fixed} = 802 \text{ miles per hour from west to east}$$

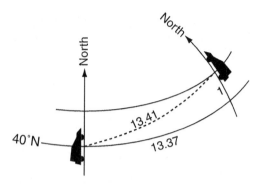

Figure A2.13. The driver of the automobile reads the odometer and sees that the car has gone exactly 1 mile in 1 minute. To the fixed observer, the vehicle moves 1 mile to the north and the planet's rotation takes the car 13.37 miles to the east in that time. This gives a speed of 13.41 miles in 1 minute, or 804 miles per hour in the nonrotating system.

That means that in 1 minute at 40°N the planet's rotation will move the vehicle 13.37 miles to the east according to the fixed observer. Now north/south:

$Vn_{fixed} = dr_n/dt + (\omega \times R)$

dr_n/dt = the speedometer reading: 60 miles per hour to the north

$\omega \times R$ = there is no rotational velocity that moves the automobile north or south

$Vn_{fixed} = 60$ miles per hour + 0

In 1 minute the vehicle will have covered a distance of 1 mile along its line of longitude. Combining 1 mile north with 13.37 miles east: $1^2 + 13.37^2 = c^2$, the fixed observer will see the car move a distance of 13.41 miles.

In these four case studies an observer in the rotating coordinate system (i.e., on the planet) perceives the speed and direction of a moving automobile differently from an observer in a fixed (nonrotating) coordinate system. Because each observer sees the exact same movement of the automobile, by using the velocity formula their two viewpoints can be reconciled.

Accounting for Centrifugal and Coriolis Forces

We have just seen how velocity can be correlated between fixed and rotating coordinating systems. This section mathematically derives acceleration from a change in velocity. Using calculus to differentiate the velocity with respect to time shows how centrifugal and Coriolis accelerations are intrinsic to unifying the accelerations of the fixed and rotating frames.

The calculus and physics that follow will show how the accelerations (and therefore the forces) are derived. You will see that Coriolis force is proportional to velocity. The faster an object moves, the greater the Coriolis force will be; if the object is still, there will be no Coriolis force. The resulting vector cross products also will show why centrifugal force is directed away from the axis of rotation, and why Coriolis force will push an object 90° from its direction of motion (left or right to be determined by the geometry of the object it is moving on).

In the previous section, velocity in a fixed coordinate system was related to velocity in a rotating coordinate system with this formula:

$$V_{fixed} = dR/dt = (\omega \times R) + dr/dt$$

Differentiating velocity with respect to time gives acceleration.

$$V_{fixed}/dt = A_{fixed} = (d\omega/dt \times R) + (\omega \times \underline{dR/dt}) + d(dr/dt)/dt$$

As defined above: $dR/dt = (\omega \times R) + dr/dt$, so

$$A_{fixed} = (d\omega/dt \times R) + [\omega \times (\omega \times R + dr/dt)] + d(dr/dt)/dt$$

$$A_{fixed} = (d\omega/dt \times R) + [\omega \times (\omega \times R)] + (\omega \times dr/dt) + \underline{d(dr/dt)/dt}$$

When differentiating with rotating coordinate axes, the **i, j, k** unit vectors that would otherwise be implied need to be treated as variables:

$$d(dr/dt)/dt = d^2r/dt^2 + (\omega \times dr/dt)$$

$$A_{fixed} = (d\omega/dt \times R) + [\omega \times (\omega \times R)] + (\omega \times dr/dt) + d^2r/dt^2$$
$$+ (\omega \times dr/dt)$$

$$A_{fixed} = \underline{d\omega/dt \times R} + [\omega \times (\omega \times R)] + 2(\omega \times dr/dt) + d^2r/dt^2$$

If angular velocity is constant (e.g., as with Earth), the first term equals zero, rewriting

$$A_{fixed} = \underline{d^2r/dt^2} + \underline{\omega \times (\omega \times R)} + \underline{2(\omega \times dr/dt)}$$

$$d^2r/dt^2 = \text{acceleration in the rotating system}$$

$$\omega \times (\omega \times R) = \text{centrifugal acceleration}$$

$$2(\omega \times dr/dt) = \text{Coriolis acceleration}$$

Now, rewriting what the fixed observer notes, to instead show the acceleration that the rotating observer sees,

$$d^2r/dt^2 = A_{fixed} - \omega \times (\omega \times R) - 2(\omega \times dr/dt)$$

Those negatives are very important; they will change the direction of the vector.

Acceleration in the rotating system is the sum of acceleration in the fixed system and centrifugal and Coriolis accelerations. Because $F = ma$, once we know the acceleration of an object we can determine the force it is subjected to if we know its mass.

Coriolis Acceleration

The value of the Coriolis acceleration, $-2(\omega \times dr/dt)$, is proportional to velocity, dr/dt. The faster the object moves, the greater the Coriolis value will be; if the object is not moving, then Coriolis will be 0. The direction of dr/dt is the direction of motion in the rotating system at the point in question. The cross product, $\omega \times dr/dt$, will have a direction perpendicular to the rotation axis and perpendicular to the direction of motion, and $-(\omega \times dr/dt)$ will have the opposite sense.

Regardless of the direction of movement, the direction of the Coriolis acceleration will be perpendicular to the direction of movement and also perpendicular to Earth's rotation axis. The latitude at which

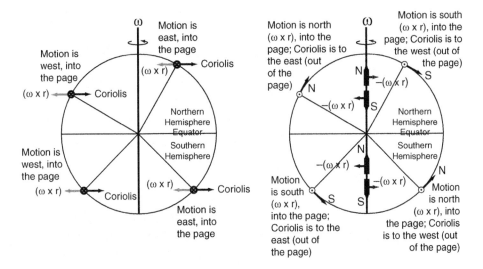

Figure A2.14. The cross product direction ($\omega \times dr/dt$) will be perpendicular to ω and to the direction of motion. The Coriolis acceleration will be in the opposite direction. *Left:* On Earth, if motion is due east or due west, the direction of Coriolis acceleration will be toward or away from the axis of rotation. *Right:* If motion is due north or due south, the direction of Coriolis acceleration will be due east or due west along Earth's surface.

movement is taking place will affect how much of the Coriolis acceleration, $-2(\omega \times dr/dt)$, will go into a turning force.

East or West Motion

Motion that is due east or due west is proceeding at 90° from the axis of rotation, so the cross product value for ($\omega \times dr/dt$) is $\sin(90°) = 1$. Nevertheless, latitude will affect how much turning influence Coriolis has, because the north or south Coriolis acceleration can be resolved into an along-surface component and into an up-down component. Because Earth is spherical, the components are a function of latitude. The along-surface component is proportional to the sine of the latitude; the vertical component is therefore proportional to the cosine. (For most purposes, the vertical component of Coriolis is meaningless—it just makes an object slightly heavier or lighter and has no effect on movement across Earth.)

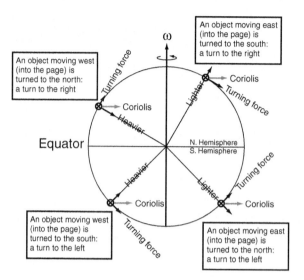

Figure A2.15. Because Earth is spherical, if motion is due east or due west, the Coriolis acceleration can be resolved into a turning component that is parallel to the surface and a vertical component that slightly changes an object's weight. The turning component is proportional to the sine of the latitude.

The Coriolis turning force is thus: $-2(\omega \times dr/dt)$ (sin latitude). It is 0 at the equator (sin 0° = 0) because all of its magnitude is perpendicular to the surface (i.e., vertical). While it would be a theoretical maximum at the poles (sin 90° = 1), east–west movement is not possible at exactly 90°.

North or South Motion

Unlike east or west movement, motion that is north or south is not perpendicular to the rotation axis (except exactly at a pole). Consequently, while the cross product, $(\omega \times dr/dt)$, gives a direction along the line of latitude (perpendicular to both the axis and the movement), its value will be reduced by the sine of the cross product angle. Because Earth is spherical, the cross product angle is the latitude. (Use the absolute value of the latitude; the direction of the resultant vector is assigned from the vector "right-hand rule," not from whether the latitude is in the Northern or the Southern Hemisphere, because hemispheres are artifacts of sphericity.)

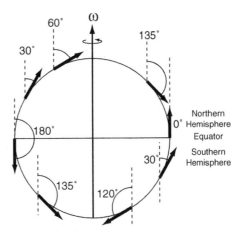

Figure A2.16.   If an object is moving due north or due south (as with these eight examples), the cross product angle between the angular velocity (i.e., the rotation axis) and the direction of motion is its latitude. The "right-hand rule" gives the direction of the resulting vector (which will be tangent to the latitude line, causing a turn to the east or to the west).

At a pole, movement is perpendicular to the axis, and the cross product is 1. At the equator, Coriolis is 0 for north or south motion because the cross product sin 0° or sin 180° = 0.

Intermediate Directions

The magnitude of the Coriolis turning force is independent of the direction in which an object is moving. Regardless of whether an object is moving in a direction other than due north, south, east, or west, the Coriolis turning force still will be a function of the sine of the latitude. Resolve the motion into north–south and east–west components (each is reduced by the sine of the latitude, although for different reasons), and then combine them back to get the Coriolis turning influence for the original motion.

The direction of the Coriolis turning force will be perpendicular to the direction of motion. The direction of the turning force is determined by the cross product angle from ($\omega \times dr/dt$), or from the trigonometric alignment of the surface. Because of Earth's spherical shape, it coincidentally works out that Coriolis in the Northern

---

Coriolis Is a Weak Force

The magnitude of the Coriolis turning acceleration, $2(\omega \times dr/dt)$, for something going 10 miles per hour (14.7 feet per second) at 40° latitude will be small in comparison to the speed of the moving object.

$$2(2\pi/24 \text{ hours})(\sin 40°) (10 \text{ miles/hour}) = 3.37 \text{ miles/hr}^2, \text{ or}$$

$$2(2\pi/86,400 \text{ seconds}) (\sin 40°) (14.7 \text{ ft/sec}) = 0.001 \text{ ft/sec}^2$$

This value is very small. Coriolis is a weak force, but it is effective at turning moving gases (e.g., atmospheric winds), moving liquids (e.g., ocean currents), or solid objects that are moving very fast.

Hemisphere will make an object turn to the right, and objects moving in the Southern Hemisphere will be turned left.

Centrifugal Acceleration

Because the ω vector points from the South Pole to the North Pole through the center of Earth, and R is a radius from the center of Earth to the point on the surface, the cross product $(\omega \times R)$ will always give a vector with a direction of due east at Earth's surface. The vector $\omega \times (\omega \times R)$ will have a cross product direction that goes from the point of interest at 90° toward Earth's axis. Therefore the negative, $-\omega \times (\omega \times R)$, will point 90° away from the rotation axis at the latitude in question.

The centrifugal acceleration always points away from the axis of rotation. Because Earth is spherical, the acceleration can be resolved into a vertical force, which works in the direction opposite to gravity, and a horizontal force. The horizontal component of centrifugal acceleration always is going to be directed north or south in a way that pushes an object toward the equator. The equatorward push will be proportional to the sine of the latitude (near a pole, 100 percent of centrifugal acceleration is directed toward the Equator; at 0° none of the centrifugal acceleration is along Earth's surface, since it is all vertical).

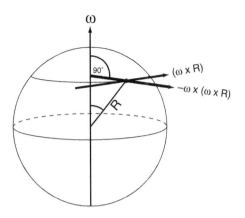

Figure A2.17. Centrifugal acceleration at a point is caused by the rotation of Earth on its axis—it is not connected with the speed or direction of an object's movement. The direction of the cross product (ω × R) will be tangent to the latitude at the point in question. The direction of the ensuing cross product, ω × (ω × R), will point toward the axis of rotation, and its negative (the direction of the centrifugal acceleration) will point 90° away from the axis, through the point in question.

Calculating Centrifugal Acceleration

At 40° N latitude on Earth, the cross product angle is 50°, and the magnitude of the centrifugal acceleration, ω × (ω × R), is going to be relatively small.

$$(2\pi/24 \text{ hours})(\sin 90°) \left[(2\pi/24 \text{ hours})(\sin 50°)(4000 \text{ miles}) \right]$$
$$= 210 \text{ miles/hr}$$

or

$$(2\pi/86{,}400 \text{ seconds})(\sin 90°) \left[(2\pi/86{,}400 \text{ seconds})(\sin 50°) \right.$$
$$\left. (21{,}120{,}000 \text{ feet}) \right] = 0.09 \text{ ft/sec}^2$$

This value can then be resolved into a horizontal component, sin(latitude), and a vertical component, cos(latitude). For comparison, the value of gravitational acceleration is about 32 ft/sec^2. The horizontal component of centrifugal acceleration is directed due south or due north along a meridian. Since a meridian is a great circle, that heading will remain true.

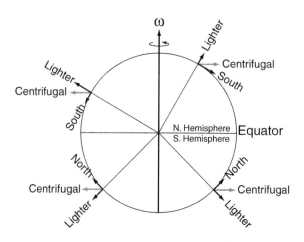

Figure A2.18. Centrifugal acceleration away from the axis is resolved into a vertical component that works against gravity and a horizontal component that pushes an object toward the equator (either north or south).

Movement Plus Coriolis Plus Centrifugal Force on a Sphere

In a simple system where an object on rotating Earth is set in motion by an initial push, it will move in three ways:

1. It will begin to travel at some speed on the great circle route in the direction in which it is pushed.

2. Coriolis force will turn the moving object to the right in the Northern Hemisphere (left in the Southern Hemisphere) of its direction of motion.

3. Centrifugal force will weakly push the still-rotating object toward the equator (south in the Northern Hemisphere, north in the Southern Hemisphere)

This describes how a mindless object that responds only to imposed forces will travel. People are not mindless, and usually they have a destination in mind when they begin moving. People continuously adjust their course by steering or following landmarks; thus they unconsciously compensate for the relatively weak Coriolis and centrifugal turning forces.

Forces from High- and Low-Pressure Centers

We have discussed mindless motion that results from an initial push. We also have discussed how people can steer or adjust their course to stay on a route. Atmospheric processes produce a different case in which a mindless object (e.g., a parcel of air) is continuously pushed by pressure forces. The parcel is not just pushed and then free to begin moving on its great circle route. Nor is the parcel being consciously steered to stay on a course. Instead, the parcel is being continuously pushed from high to low pressure, and the direction of the pushing is constantly changing.

The balances between pressure and Coriolis forces and the orbiting that those forces compel will keep the parcel from moving blindly around Earth along a great circle path. An object (for example, a cannonball) that is given only an initial impulse will follow a predictable

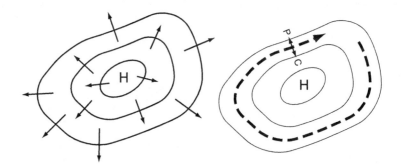

Figure A2.19. *Left:* A parcel of air will be pushed away from areas of higher pressure toward areas with lower pressure. *Right:* When the pressure force sets the air in motion, Coriolis will turn it to the right in the Northern Hemisphere. The turning from the Coriolis force will balance the pressure force to set up a rotating flow. As it orbits the high-pressure center, the parcel of air is continuously subjected to a pressure force compelling movement, but the direction that pressure wants the air to go is constantly changing as the orbital position of the parcel changes.

route. In contrast, an object that is ceaselessly pushed in changing directions will follow a path that accommodates all the forces imposed on it.

Don't Get Taken by Deceptive Analogical Explanations

Coriolis is not an easy thing to understand or to explain. Sometimes to save time, explanations make use of analogies to simplify this difficult concept. There are two analogies that are commonly invoked. Explanations that use flying cannonballs or missiles moving along while Earth rotates beneath them are almost always not correct. And usually they don't make sense once they are questioned. Also, it is sometimes said that the clock-respective orientation of the rotation affects Coriolis direction. Explanations that say one hemisphere rotates clockwise and the other goes counterclockwise and that is the reason for why Coriolis is to the left or to the right are useful but are not mathematically correct.

Shooting Southward

The cannon shooting over Earth's northern hemisphere while the planet mindlessly spins along is a false example of the Coriolis effect. The example almost always takes the form of a cannon near the North Pole shooting due south, toward the equator. It is said that as the cannonball flies along, Earth rotates underneath it, and thus the shot ends up to the west of the place at which it was aimed. After that (which is wrong anyway), the explanation falls completely apart.

If the cannon is shot north from low latitudes toward the pole, the same thing should happen, right? If the explanation is correct, Earth should rotate beneath the north-flying ball, and it will land west of its target. But that isn't what happens in the real world: the ball turns to the right, not to the left. This explanation also fails to account for why a cannonball shot due east will turn to the right, toward the equator, or why a cannonball shot due west will turn right, toward the North Pole. This explanation is completely flawed and is unredeemable.

Clock-Respective Spins

The Northern Hemisphere experiences counterclockwise spin, from west to east. The Southern Hemisphere experiences clockwise spin,

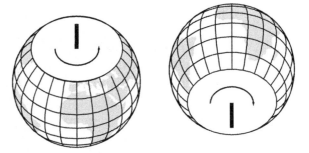

Figure A2.20. Although every place on Earth moves from west to east, it is sometimes said that the Northern Hemisphere rotates counterclockwise and the Southern Hemisphere rotates clockwise.

from west to east. Places exactly on the equator do not spin at all; they just move west to east. There is an analogy to the counterclockwise rotation we saw earlier with the rotating disk when rotating observers see the ball turn to the right of its path. If the disk instead rotated clockwise, the observers on the disk would see the ball turn to the left.

Thus Coriolis force is said to make moving things turn to the right of their path when they are in the Northern Hemisphere. In the Southern Hemisphere, Coriolis force makes moving things turn to the left. While this is a literally true description, it is an incorrect explanation. It is not a clockwise or a counterclockwise spin that is making the moving object turn differently in the northern and southern sections of spherical Earth.

Every place on Earth rotates west to east. When viewed from above

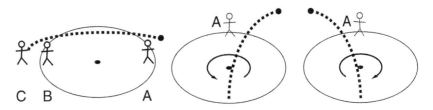

Figure A2.21. *Left:* In the simple case from the beginning of the chapter, Chris threw a ball over Betty's head past Andrew. *Center:* If the disk rotated counterclockwise, we noted that Betty would see the ball turn to the right and miss Andrew. *Right:* If the disk spun clockwise instead, Betty would see the ball miss Andrew because it turned to the left.

the North Pole, every place moves counterclockwise. The angular velocity of every place on Earth is identical. Changing viewpoints for different hemispheres might make for a clever explanation, but it doesn't change the fact that Earth's rotation is turning every place at the same angle in the same direction in the same amount of time as every other place. It is the spherical shape of Earth and the effects of geometry on how Coriolis forces are resolved that constitute the reason for right or left turns in different hemispheres (and of no turning at the equator). It is a circumstance of the geometric properties of a sphere that allows this shortcut to work even if the explanation is not precisely correct.

Index

CPSIA information can be obtained at www.ICGtesting.com
Printed in the USA
BVOW09s1646011115

424802BV00009B/18/P